Wireless Sensor Networks - Design, Applications and Challenges

Edited by Jaydip Sen, Mingqiang Yi, Fenglei Niu and Hao Wu

Published in London, United Kingdom

Wireless Sensor Networks - Design, Applications and Challenges
http://dx.doi.org/10.5772/intechopen.104033
Edited by Jaydip Sen, Mingqiang Yi, Fenglei Niu and Hao Wu

Contributors
Poornima G. Miathali, Rakesh Chandra Gangwar, Roohi Singh, Sumana Naskar, Christian Osueke, Segun Adebayo, Adefemi Adekunle, Reuben S. Diarah, Adedayo Banji Aaron, Olaluyi Olawale Joshua, Radu Setnescu, Eduard-Marius Lungulescu, Jaydip Sen, Deqi Chen, Fenglei Niu, Hao Wu, Qianlong Zuo, Haidong Liu, Mingqiang Yi

First published in London, United Kingdom, 2023 by IntechOpen
IntechOpen is the global imprint of INTECHOPEN LIMITED, registered in England and Wales, registration number: 11086078, 5 Princes Gate Court, London, SW7 2QJ, United Kingdom

British Library Cataloguing-in-Publication Data
A catalogue record for this book is available from the British Library

Additional hard and PDF copies can be obtained from orders@intechopen.com

Wireless Sensor Networks - Design, Applications and Challenges
Edited by Jaydip Sen, Mingqiang Yi, Fenglei Niu and Hao Wu
p. cm.
Print ISBN 978-1-80355-453-2
Online ISBN 978-1-80355-454-9
eBook (PDF) ISBN 978-1-80355-455-6

Meet the editors

Jaydip Sen is a professor in the Department of Data Science, Praxis Business School, Kolkata, India. His research areas include security and privacy issues in wireless ad hoc and sensor networks, intrusion detection systems, machine learning, deep learning, and artificial intelligence in the financial domain. He has published more than 200 papers in reputed indexed journals, refereed international conference proceedings, and 18 book chapters. He has authored three books and edited ten volumes. He is the editor of *Knowledge Decision Support Systems in Finance* and serves on the technical program committees of several high-ranked international conferences of the Institute of Electrical and Electronics Engineers (IEEE) and the Association for Computing Machinery (ACM). He has contributed to several standardization efforts of the IEEE, including the 802.16m standards and the 3GPP LTE standards. Prof. Sen has been listed among the top 2% of scientists in the world by Stanford University for the last four consecutive years (2019–2022).

Dr. Mingqiang Yi has been CEO of Microfluidic Foundry since 2011. He also served as the principal investigator for a project funded by the National Health Institute in 2014–2015. From 2009 to 2011, Dr. Yi was the process engineering manager for Alphabet Energy, Inc., where he was responsible for the development of silicon nanowire-based thermoelectric modules. From 2005 to 2008, he was the project leader for RheoSense, Inc., where he was responsible for the development of microfluidic chip-based viscometers. From 2001 to 2002, Dr. Yi was a MEMS scientist at Los Gatos Research, Inc., responsible for the development of high-pressure electrokinetic micropumps. From 2002 to 2005, he was a postdoctoral fellow at the University of California, Berkeley, USA, where he developed microfluidic sensors for detecting DNA hybridization and enzymatic reaction products as well as for protein analysis. Dr. Yi obtained his Ph.D. in Mechanical Engineering from the University of Pennsylvania, USA, in 2001, and a BEng and MEng from Tsinghua University in 1994 and 1997, respectively. He has a broad background in micro/nanotechnology, with specific training and expertise in microfluidics.

Fenglei Niu is a pioneering researcher in nuclear thermal hydraulics. He is former dean and professor of the School of Nuclear Science and Engineering, North China Electric Power University and a member of the American Nuclear Society. Dr. Nie has multiple registered patents to his credit.

Dr. Hao Wu is an assistant professor at the School of Nuclear Science and Engineering, North China Electric Power University, Beijing, China. He received his Ph.D. from Tsinghua University, China. He worked for the Institute of Nuclear and New Energy Technology of Tsinghua University as a postdoctoral researcher. He has thirty publications to his credit, including fourteen research articles as the first author in international journals such as *Annals of Nuclear Energy, International Journal of Heat and Mass Transfer,* and *Chemical Engineering Science*. His research interests are thermal hydraulics in advanced nuclear reactors, including solid-phase oxygen control, numerical oxygen mass and heat transfer simulations in the lead–bismuth eutectic, deep learning applications in nuclear engineering, particle-scale thermal radiation and continuum model in the pebble-bed reactor and CFD-DEM modeling in dense fluid-particle dynamics.

Contents

Preface

Wireless Sensor Networks (WSNs) are networks of spatially distributed autonomous sensor nodes that monitor a variety of ambient conditions, physical characteristics, or events and transmit the data they collect to a central processing hub or server. These networks have attracted a lot of attention because of the diverse variety of applications they might be used in, including surveillance, health care, agriculture, industrial automation, and environmental monitoring. Some key aspects of a typical WSN are highlighted in the paragraphs that follow.

Sensor Nodes: The basic components of WSNs are sensor nodes. They are made up of a variety of parts, including memory, communication interfaces, processing units, sensing units to measure physical parameters like temperature, humidity, light, and so on, and processing units to process acquired data.

Communication: Wireless communication techniques are often used over radio frequencies for communication between sensor nodes. Data may have to travel via several nodes in the communication process to reach the center or the sink node.

Data Aggregation: Due to the large number of nodes in a network, it is essential to minimize the amount of data transmitted to conserve energy and reduce network congestion. Data aggregation techniques help to combine and summarize data at intermediate nodes before transmitting it to a central node.

Energy Constraints: One of the significant challenges in WSNs is the limited energy resources of individual sensor nodes. Many nodes are often deployed in remote or hard-to-reach areas, making battery replacement or recharging impractical. Energy-efficient protocols and algorithms are crucial to prolong the network's operational lifetime.

Routing Protocols: The routing protocols used in WSNs need to be power efficient so that data transmissions from source nodes to the destination nodes occur with minimal energy consumption. Based on network features and application requirements, many routing strategies, including hierarchical routing, data-centric routing, and geographic routing, are used.

Localization: For applications that need spatial awareness, such as target tracking or monitoring environmental changes, accurate localization of sensor nodes is crucial. For localization, usually several methods are utilized, such as global positioning systems (GPS), triangulation, and distance estimates.

Security: Since WSNs very often involve processing and communication of sensitive data, security of communication and privacy of data is a major challenge for these networks. These networks must deploy protocols for data encryption, authentication, and secure communication because sensor nodes can be physically attacked and eavesdropped on.

Scalability: Because many applications of WSNs involve a very large number of nodes, scalability must be guaranteed in the network design without compromising computing and communication efficiencies. Effective management of large-scale networks most often requires the use of hierarchical network design and clustering approaches.

Applications: WSNs find applications in diverse fields, including environmental monitoring (e.g., tracking pollution levels), agriculture (e.g., monitoring soil moisture), health care (e.g., patient monitoring), smart cities (e.g., traffic monitoring), and industrial automation (e.g., monitoring equipment conditions).

Overall, WSNs have the potential to revolutionize various industries by providing real-time data insights, enabling automation, and improving decision-making processes. Despite their numerous applications in various domains, WSNs face several challenges. These challenges include designing energy-efficient routing protocols, enhancing the robustness of networks, ensuring security in communication and privacy of data, and minimizing the delay in real-time communications as far as possible. These challenges require ongoing research to develop better algorithms, protocols, and hardware solutions.

The chapters of this volume deal with various research issues and applications of WSNs, including key management, routing, machine learning-based applications in WSNs, temperature sensors and their different applications, automatic fault detection in WSNs, and more.

In Chapter 1, "Introductory Chapter: Temperature Sensing and Advanced Applications", Mingqiang Yi and Fenglei Niu highlight various temperature sensors and their applications in the real world. The authors briefly point out the recent advances in smart temperature sensing including control of the pandemic, point-of-care health care, and monitoring of remote and geographically challenging environments.

In Chapter 2, "A Survey of Cryptography and Key Management Schemes for Wireless Sensor Networks", Jaydip Sen presents a comprehensive overview of various cryptographic and key management schemes for WSNs. The author first discusses symmetric key and public key cryptographic approaches and highlights their merits and demerits. The key management schemes are then further categorized into three groups: network-topology-based, deterministic key distribution-based, and probabilistic key distribution-based. The author critically analyzes several propositions in the current literature under each category and highlights some open research questions for future exploration.

In Chapter 3, "Machine Learning Algorithms from Wireless Sensor Network's Perspective", Rakesh Chandra Gangwar and Roohi Singh present an empirical study of architecture, applications, and challenges in WSNs. The authors highlight the critical importance of energy efficiency, security, delay tolerance, localization, and data aggregation in sensor networks. Further, the authors present a taxonomy of machine learning algorithms for dynamic adaptations of WSNs in response to the changes in their working environment.

In Chapter 4, "Efficient Machine Learning Classifier for Fault Detection in Wireless Sensor Networks" Poornima G. Miathali presents a study on the faults in WSNs caused by denial of service (DoS), probe, remote-to-local (R2L), and user-to-root (U2R) attacks. The KDD Cup 99 dataset is used with 41 features in total, and for each fault type, the relevant features are identified using a recursive feature elimination method. Different classifiers are trained on the training dataset and their performances are evaluated on the test sample. The author observes that the random forest model performed the best among all classifiers.

In Chapter 5, "Wireless Sensor Networks Challenges and Solutions", Sumana Naskar summarizes various important aspects of WSNs, including their architecture, applications, challenges, and potential solutions. The author identifies some of the research directions as well, which include sensor node deployment, relay node and cluster head node selection, energy optimization, security and privacy issues, dela minimization, and fault tolerance.

In Chapter 6, "Novel PTC Composites for Temperature Sensors (and Related Applications)", Radu Setnescu and Eduard-Marius Lungulescu present an overview of conductive polymer composites, focusing on those that exhibit abrupt changes in resistivity with variations in temperature, known as the positive temperature coefficient (PTC) and negative temperature coefficient (NTC) effects. The authors further analyze the factors that affect the quality of polymer composites such as polymeric matrix, conductive filters, and treatments. Temperature sensors based on PTC composites are also discussed as illustrative examples.

In Chapter 7, "Current Status and State-of-Art Developments in Temperature Sensor Technology" Deqi Chen et al. argue that while the traditional temperature sensors currently in use are simple and reliable, their performance in challenging and harsh environments is questionable. Accordingly, the authors discuss various types of advanced temperature sensors that offer unique benefits and applications compared to their traditional counterparts. Some specific examples are discussed in the chapter, including acoustic temperature sensors for monitoring high-temperature boilers, fiber-optic temperature sensors for long-distance cables, and micro- and nano-scale sensors for monitoring microstructures.

In Chapter 8, "Types of Temperature Sensors", Reuben S. Diarah et al. discuss various types of temperature sensors and their applications. In particular, the authors present three types of temperature sensors: thermometers, resistance temperature detectors, and thermocouples. Further, the authors delve into the working principles, sizes, and ranges of temperature, and applications for each category of sensors.

I am hopeful that this volume will prove to be a valuable resource for researchers, engineers, doctoral students, and faculty members from graduate schools and universities who are engaged in the design, implementation, and deployment of WSNs. The content presented in the chapters addresses advanced topics within the realm of WSNs, making it essential for readers to possess a foundational understanding of the covered subjects. This book is tailored to individuals with the necessary background knowledge to fully grasp and benefit from its contents.

I extend my heartfelt gratitude to all the esteemed authors for their invaluable contributions. Their collaborative efforts and scholarly input have paved the way for the successful publication of this book. I would like to express my sincere appreciation to

Author Service Manager Ms. Paula Gavran at IntechOpen for her unwavering support, patience, and cooperation throughout the publication process. My thanks also go to Commissioning Editor Ms. Danijela Vladika at IntechOpen for entrusting me with the significant responsibility of editing this academic volume.

I am indebted to my colleagues and graduate students at the Praxis Business School, Kolkata, India, for their constant support and cooperation. My family members have been a constant source of inspiration and motivation, and I dedicate this volume to the memory of my beloved sister, Ms. Nabanita Sen, who sadly succumbed to cancer on September 27, 2021. She was my steadfast pillar of strength. Lastly, I am profoundly grateful for the unwavering encouragement and motivation from my wife, Ms. Nalanda Sen, my daughter, Ms. Ritabrata Sen, and my mother, Ms. Krishna Sen. Their support and inspiration have been the driving force behind the publication of this book.

Jaydip Sen
Professor,
Department of Data Science,
Praxis Business School,
Kolkata, India

Mingqiang Yi and Fenglei Niu
Microfluidic Foundry,
San Pablo, California, USA

Hao Wu
School of Nuclear Science and Engineering,
North China Electric Power University,
Beijing, China

Chapter 1

Introductory Chapter: Temperature Sensing and Advanced Applications

Mingqiang Yi and Fenglei Niu

1. Introduction

Temperature measurements are vital to our life and various activities. Temperature sensors are devices that measure the target temperature and convert it into an electrical signal. Many temperature sensors and measurement technologies have been developed and provided in industry and laboratories, including temperature probesused for immersion temperature measurement in liquids, air, gas, or wells in solid material, and surface sensors such as fast-response thermocouples mounted in a polyimide carrier for easily measuring surface temperatures or heat flux. Fiber optic monitors and fiber optic sensors have been designed to work more reliably than traditional sensors due to their resistance to outside influence such as microwaves, electromagnetic interference, and radio frequency interference. There are also non-contact temperature measurement solutions for remote measurement or human body temperature measurement. Recently, the smart temperature sensing has greatly advanced the applications such as pandemic mitigation, point-of-care healthcare, and harsh environment monitoring [1, 2].

There are many different types of temperature sensors, each with its own advantages and disadvantages [3, 4]. Some of the most common types of temperature sensors include:

- Thermocouples: Thermocouples are the most common type of temperature sensor. They are relatively inexpensive and easy to use. However, they are not very accurate and can be affected by environmental factors such as humidity.

- Resistance temperature detectors (RTDs): RTDs are more accurate than thermocouples, but they are also more expensive. RTDs are typically made of platinum or nickel, and their resistance changes with temperature.

- Semiconductor temperature sensors: Semiconductor temperature sensors are the most accurate type of temperature sensor. They are also very small and can be integrated into electronic devices. However, semiconductor temperature sensors are also the most expensive type of temperature sensor.

The advanced applications of temperature sensing are constantly evolving. Some of the most exciting new applications include:

IntechOpen

- Wearable temperature sensors: Wearable temperature sensors are being used to monitor the health of patients and athletes. These sensors can be used to track body temperature, heart rate, and other vital signs.

- Surface temperature sensors: These are used to measure the temperature of any flat, curved, or moving surface, including those of self-adhesive, cement-on, bolt-on, magnetic mount, or handheld surface temperature sensors.

- Temperature wire sensors: These simple, temperature-sensing element and extension wire combinations have a wide array of applications and are available in thermocouple, RTD, and thermistor technologies.

- Temperature probes: These are used for immersion temperature measurement in liquid, air, gas, or wells in solid material, including probe styles in thermocouple, RTD, thermistor, and IC technologies.

- Fiber optic temperature measurement: Fiber optic monitors and fiber optic sensors are designed to work more reliably than traditional sensors due to their resistance to outside influence such as microwaves, electromagnetic interference, and radio frequency interference.

- Hybrid temperature sensors: These perform high accuracy non-invasive hybrid temperature measurement for various fluids that eliminate the costly process downtime and require no welding or drilling.

- Internet of Things (IoT) temperature sensors: IoT temperature sensors are being used to monitor the temperature of buildings, machines, and other assets. These sensors can be used to detect problems early on and prevent damage.

- Smart agriculture: Smart agriculture is using temperature sensors to monitor the temperature of soil, water, and crops. This information can be used to optimize crop yields and reduce water usage.

- Climate change research: Temperature sensors are being used to monitor the Earth's climate. This information is being used to study the effects of climate change and to develop solutions to mitigate its effects.

Temperature sensing is a versatile technology with a wide range of applications. As the technology continues to evolve, we can expect to see even more advanced applications in the future. Below are some additional examples of advanced applications of temperature sensing:

- In the medical field, temperature sensors are used to monitor the temperature of patients in critical care units. This information can be used to detect infections and other medical problems early on.

- In the food industry, temperature sensors are used to monitor the temperature of food during processing and storage. This information can be used to ensure that food is safe to eat.

- In the manufacturing industry, temperature sensors are used to monitor the temperature of machinery. This information can be used to prevent equipment failure and to improve product quality.

Temperature sensors are important devices that are used in a wide variety of applications. Recent advances in temperature sensor technology have led to the development of new types of temperature sensors with improved sensitivity, accuracy, and range. Temperature sensing is a powerful tool that can be used to improve our lives in many ways. As the technology continues to evolve, we can expect to see even more innovative applications in the future.

Author details

Mingqiang Yi* and Fenglei Niu
Microfluidic Foundry, San Pablo, California, USA

*Address all correspondence to: myi@microfluidicfoundry.com

References

[1] Bhar I, Mandal N. A review on advanced wireless passive temperature sensors. Measurement. 2022;**187**:110255

[2] Dinh T, Phan HP, Qamar A, Woodfield P, Nguyen NT, Dao DV. Thermoresistive effect for advanced thermal sensors: Fundamentals, design considerations, and applications. Journal of Microelectromechanical Systems. 2017;**26**(5):966-986

[3] Childs PR, Greenwood JR, Long CA. Review of temperature measurement. Review of Scientific Instruments. 2000;**71**(8):2959-2978

[4] Temperature Measurement. https://www.omega.com/en-us/c/temperature-measurement

Chapter 2

A Survey of Cryptography and Key Management Schemes for Wireless Sensor Networks

Jaydip Sen

Abstract

Wireless sensor networks (WSNs) are made up of a large number of tiny sensors, which can sense, analyze, and communicate information about the outside world. These networks play a significant role in a broad range of fields, from crucial military surveillance applications to monitoring building security. Key management in WSNs is a critical task. While the security and integrity of messages communicated through these networks and the authenticity of the nodes are dependent on the robustness of the key management schemes, designing an efficient key generation, distribution, and revocation scheme is quite challenging. While resource-constrained sensor nodes should not be exposed to computationally demanding asymmetric key algorithms, the use of symmetric key-based systems leaves the entire network vulnerable to several attacks. This chapter provides a comprehensive survey of several well-known cryptographic mechanisms and key management schemes for WSNs.

Keywords: wireless sensor network (WSN), public key, symmetric key, key management, cryptography, key distribution, random key distribution, security

1. Introduction

Wireless sensor networks (WSNs) are made up of a large number of tiny sensors, which can sense, analyze, and communicate information about the outside world. These networks play a significant role in a broad range of fields, from crucial military surveillance applications to monitoring building security [1]. In these networks, a sizable number of sensor nodes are placed throughout a big field, where the operational environment is frequently hostile or severe, to monitor it. However, because of their low processing speed, little memory, and insufficient energy, WSN nodes face significant resource limitations. Hence, these networks need to include security features to protect against attacks like physical tampering, node capture, denial of service, eavesdropping, etc. as they are typically placed in distant locations and left unattended.

Unfortunately, resource-constrained sensor nodes cannot implement typical security measures because of their large overhead. Researchers in WSN security have put out many security protocols that are tailored to these networks' resource limitations.

Researchers in WSN security have proposed several protocols for secure and efficient routing [2–5], securely aggregating data for protecting data privacy [6–11], etc.

Since WSN architectures are mostly decentralized, and due to the lack of any infrastructure, security procedures used in WSNs need also to incorporate cooperation among the nodes along with addressing more security challenges like secure routing and aggregation of data. In the real-world deployment scenario, WSNs cannot be a priori taken to be reliable. To address the issues that standard cryptographic algorithms are unable to address, researchers have concentrated on developing a sensor trust model [12–19].

Vulnerability to physical attacks is a significant concern in WSNs since the sensor nodes are typically unattended and physically unsafe. There are several ideas in the literature for protecting sensor nodes from physical attack [20–29].

The choice of the cryptographic scheme and the key distribution and management protocol for a WSN is an extremely critical decision as the entire security of the network is based on these schemes. However, designing a computationally efficient yet highly secure key management scheme is a challenging task. While these resource-constrained sensor nodes should not be exposed to computationally demanding public key-based algorithms, the use of symmetric key cryptography leaves the network vulnerable to several attacks. This chapter provides a comprehensive survey and a comparative analysis of various cryptographic mechanisms and key management schemes in the current literature.

The rest of the chapter is organized as follows. Section 2 presents different cryptographic schemes used in WSNs including the public key and the symmetric key-based algorithms and systems. Section 3 discusses several key management schemes including the network architecture-based protocols and deterministic, and probabilistic key distribution mechanisms. Finally, Section 4 concludes the chapter and highlights some future research directions.

2. Cryptographic schemes for WSNs

In WSNs, choosing the best cryptographic technique is essential since cryptography provides all security functions. The code size, data size, processing time, and power consumption of cryptographic techniques used in WSNs should all be taken into consideration together with the sensor node limits. We concentrate on the choice of cryptography in WSNs in this section. We first discuss public key cryptography, then delve into systems that use symmetric keys for their cryptographic functions.

2.1 Public key cryptographic mechanisms in WSNs

Many experts think that public key protocols such as the Diffie-Hellman key exchange [30] or RSA [31], should not be used in WSNs because of the code complexity, data size, processing time, and power consumption these algorithms involve.

A single security operation typically requires dozens or even millions of multiplication instructions, which makes public key methods like RSA computationally demanding. Furthermore, the number of CPU cycles needed to execute an instruction for the multiplication operation is a critical factor in determining a microprocessor's efficiency for a public key method [32].

In resource-constrained wireless devices, Brown et al. discovered that public key methods like RSA typically take some minutes to execute cryptographic operations such as encryption and decryption. This is a long enough time for an adversary to launch denial of service (DoS) attacks [33]. Carman et al. observed that a basic 128-bit operation of multiplication often requires thousands of nano-joules from a microprocessor [32].

As opposed to public key methods, the algorithms of hash functions and symmetric keys involve substantially lower processing overhead. An AES block of 128-bit size typically consumes an energy of 0.104 mJ, which is substantially lower than the anticipated energy consumption for a 1024-bit block when utilizing RSA on the MC68328 DragonBall CPU [32].

By employing the appropriate choice of parameters in the algorithms and optimized approaches that consume lower power for execution, research has demonstrated that it is possible to deploy public key-based protocols in WSNs [34–36]. Elliptic Curve Cryptography (ECC) [37, 38], Ntru-Encrypt [39], RSA [31], and Rabin's Scheme [40] are some of the public key algorithms that have been studied for this purpose. The RSA and ECC algorithms are the subjects of most studies in the literature. ECC is appealing because it is highly secure even with smaller keys. Hence, the use of ECC decreases the requirement of processing and transmission costs. While RSA with 1024-bit keys offers a degree of security that is currently acceptable for many applications, the same level of security is achieved using ECC with a 160-bit key (ECC-160) [41]. As per the new recommendation, a key size of 2048 bits is used in the RSA protocol as the minimum size of the key. This is similar to the 224-bit ECC protocol [42].

On an Atmel ATmega128 CPU, Wander et al. evaluated the amount of energy required in RSA and ECC protocols for authentication and key exchange [36]. The Elliptic Curve Digital Signature Algorithm (ECDSA) generates and verifies the ECC-based signature [43]. The handshake in the secured socket layer (SSL) requires two entities: a client that initiates the session, and a server that responds to the request [44]. The key exchange scheme is a more compact form of this handshake. Each sensor in the WSN is presumed to have a certificate that has been signed using the private key of the trusted authority. The two parties validate their respective certificates during the handshake phase and agree on the session key that will be used for communication. The findings indicate that compared to RSA signatures, ECDSA signatures are much less expensive. Additionally, on the server side, the ECC protocol has superior performance, while the RSA protocol performs better on the client side. However, the two protocols do not exhibit any significant difference in the power requirement in carrying out the key exchange operation. Additionally, as the key size grows, ECC outperforms RSA in terms of relative performance.

The use of encryption operations in RSA and ECC on Mica2 motes demonstrated the viability of the use of public key protocols in WSNs [45]. The design of the TinyPK system proposed by Watro et al. uses the TinyOS development environment to build the RSA system on Mica2 motes [46]. The authors have shown that this technique effectively implements authentication and key agreement protocol in sensor nodes with limited resources. Another ECC-based technique called TinyECC [47] has been created and put into use on Mica2. Malan et al. also carried out similar work using ECC on Mica2 [45]. A single symmetric key was distributed via ECC for the TinySec module's link-layer encryption.

While sensor nodes could be able to perform public key cryptography, the cost of private key operations remains high. In some cases, the [35, 45] assumptions might not be true. For instance, [35, 45] solely focused on the public key activities, presuming a base station or outside party would handle the private key operations. The operation time of the public key may be made to be very quick by choosing the right parameters, for instance, by utilizing the tiny number $e = 2^{16} + 1$ as the public key, while the operation time of the private key remains constant. Several public key operations are not available in this framework due to the restriction of operations using private keys exclusively at a base station. Peer-to-peer authentication and secure data are two examples of such services.

2.2 Symmetric key cryptography in WSNs

As symmetric key cryptography approaches involve less computational overhead than public key cryptographic mechanisms, most research studies for WSNs concentrate on their utilization. A single shared key between the two communicating hosts is employed by symmetric key cryptographic techniques and is used for both encryption and decryption. But efficiently and securely distributing a common key to two nodes for secure communication is a significant barrier to the widespread use of symmetric key encryption. Given that it might not always be possible to pre-distribute the key, this is a challenging topic.

Five well-known encryption techniques were tested on six different microprocessors, with word sizes ranging from 8 bits (Atmel AVR), 16 bits (Mitsubishi M16C), and 32 bits (StrongARM, XScale) in [48]. These included RC4 [49], RC5 [50], IDEA [49], SHA-256 [51], and MD5 [49, 52]. For each algorithm and platform, execution time and code memory size were assessed. The studies showed that each encryption class and architectural class had a consistent cryptographic cost. While support for the Instruction Set Architecture (ISA) is only confined to certain impacts on specific protocols, the influence of caches was minimal. Additionally, hashing techniques (like MD5 and SHA-1) are found to consume more resources in comparison to RC4 and IDEA encryption algorithms.

Law et al. studied the performance of the RC5 and TEA symmetric key algorithms in [53]. On the MSP430F149 from IAR Systems, six additional block ciphers are also assessed [53]. These block ciphers are Rijndael, Camellia, KASUMI, RC6, and RC5. The benchmarking criteria were CPU cycles, data RAM, and code.

For WSN security services to be provided, the proper cryptography mechanism for sensor nodes must be chosen. The capability of the sensor nodes for calculation and transmission, however, determines the outcome. Hardware design and encryption algorithms are both active areas of study.

As mentioned earlier, studies have observed the viability of public key-based protocols in WSNs even if they have higher resource requirements. Private key operations can still not be completed in a sensor node due to the high computational and energy costs involved. Further research is needed on the use of operations using symmetric keys shared among the nodes in a WSN. In terms of speed and low energy consumption, symmetric key cryptography is preferable to public key cryptography. However, key distribution methods using shared symmetric keys are not flawless. Designing effective and adaptable key distribution strategies is necessary. To meet the growing demands on computing and communication in sensor nodes, it is also anticipated that stronger motes will need to be developed.

3. WSN key management protocols

Key management has gotten the most attention from researchers studying WSN security. A crucial strategy for ensuring network services and application security in

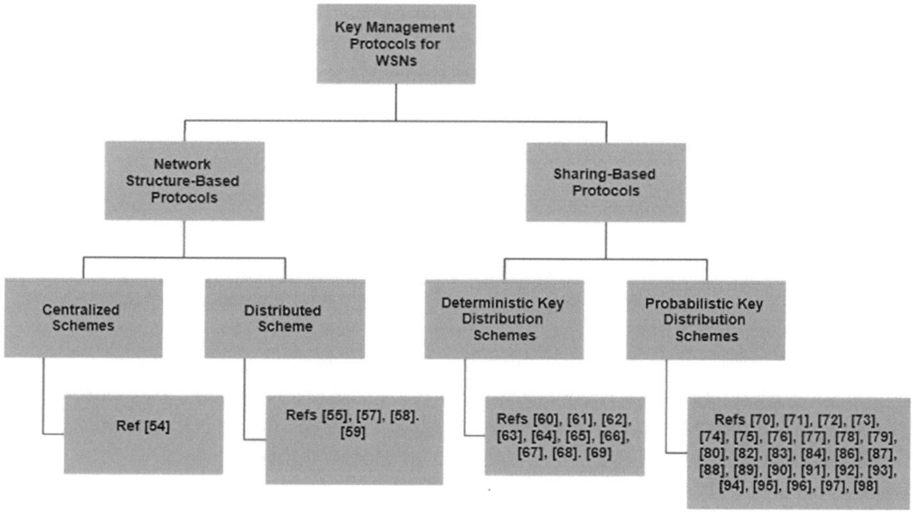

Figure 1.
The classification of key management schemes for WSNs.

WSNs is key generation, storage, and distribution. Establishing keys for the nodes efficiently and securely is the main objective of a key management scheme. The key management system should allow for network node insertion and revocation. These methods must be extremely lightweight even for a large-scale network, due to the power and memory constraints of the nodes. Due to their high computational overhead, the public key cryptographic approaches do not find many applications in WSNs. Most of the protocols for key management are based on the use of shared keys. **Figure 1** depicts a classification of the currently existing key management schemes for WSNs. The works discussed in this section are from impactful publications in the WSN literature from 2000 to 2022.

3.1 Network architecture-based key management schemes

The distribution or centralization of the key management task depends on the underlying architecture of the network. The production, distribution, and revocation of keys are all under the control of one entity under a centralized key management scheme. The name of this organization is Key Distribution Centre (KDC). The LKHW method is a protocol for WSNs that uses a single key distribution in a centralized manner [54]. The basis of the LKHW protocol is the hierarchy of logical keys. The hierarchy of the keys leads to a tree structure in which the base station is at the root position of the tree. The base station plays the role of the KDC in the network. This scheme's sole point of failure is its biggest flaw. The whole network's security will be compromised if the central controller malfunctions. Another problem is that it cannot be scaled. Additionally, it does not offer data authentication. Different controllers are used in the distributed key management protocols to manage key-related tasks. These protocols enable higher scalability and do not have a single point of failure problem. The majority of key management techniques that have been studied so far are dispersed in nature.

To secure the sensor network, Qin et al. [55] proposed an approach that involves building an AVL tree [56] for key management along with the use of elliptic curve cryptography (ECC) [36]. The AVL tree stores each node's public key and the identifier of its neighboring nodes. The scheme is efficient from several aspects, including processing overhead, memory space requirement, and overhead of communication. *Elliptic curve pallier encryption* (ECPE), a cryptographic technology, is also used in this strategy to defend against numerous security risks. Another element of this strategy was constantly updated keys.

A scheme proposed by Swaminathan and Vivekanandan [57] uses the topology of a wireless network and creates a structure aggregating several distributed spanning trees (DSTs). The proposed scheme, known as the *efficient low-cost key generation mechanism* (ELWKM), is found to be involving low overhead in computation and memory requirement.

An efficient public key cryptography-based strategy was presented by Chen et al. [58]. The scheme combined the Merkle hash tree, the Bloom filter, and several other encryption and decryption techniques. The elliptic curve discrete logarithm issue makes use of key threshold theory to create a key management system.

For clustered WSNs, Yao et al. presented a key management method known as the *local key hierarchy* (LKH++) [59]. A dynamically constructed tree is used for storing the keys in the nodes of the network. For secure communication among a group of nodes in a cluster, the keys are used for encryption and authentication. The sink node i.e., the base station stores and manages the tree. When needed for the network, this method regenerates and rekeys the keys. The LKH++ scheme provides a WSN with increased robustness against several attacks.

These methods, which can be classified as deterministic or probabilistic, are covered further in this section.

3.2 Sharing-based key management schemes

The likelihood of the availability of a shared key between any two randomly chosen nodes in a WSN is used as a basis for the classification of the key management techniques. The essential management strategies might be either deterministic or probabilistic depending on this likelihood.

3.2.1 Deterministic key distribution schemes

Zhu et al. proposed a protocol for key distribution protocol in WSNs [60, 61]. The scheme, known as the *localized encryption and authentication protocol* (LEAP), is based on cryptographic operations using shared keys among the nodes. Depending on the security needs of each packet, it employs a separate keying scheme. Each node is assigned one of four different types of keys: (i) a unique pre-distributed key shared between the nodes and the sink node, (ii) a set of keys shared among the nodes of the network, (iii) keys shared among neighboring node pairs, and (iv) a shared key among all members of a cluster. Peer-to-peer communication is secured using the pair-wise keys shared with nearby nodes, and local broadcast is secured using the cluster key.

The time needed to launch an attack on a node is longer than that needed for the network to build. A node will be able to discover all its intermediate neighbors during this period. Each node is deployed with a shared initial key already loaded. A master key is generated for each node based on their shared key and the individual identity of the node. Then, sensor nodes communicate by exchanging hello messages. The hello messages are verified by the recipients (the neighbor's master key may be calculated because the shared key and identification are known). Based on their master keys, the nodes then compute a shared key. After the setup, the common key is deleted in every node, and it is assumed that no node has been hacked thus far. Injecting bogus data or decoding messages sent earlier is now very difficult as no attacker can get access to the shared key. No node may afterward fabricate the master key of another node, either. This establishes the shared keys for all node pairs among all neighboring nodes located nearby. A node creates the cluster key after the keys for the node pairs are generated. With the help of the shared key between a node pair, the cluster is derived and the cluster key is delivered in an encrypted form to all the neighboring nodes. The group key is installed in the nodes a priori, and it is revoked and regenerated as soon as a compromised node is found. In a crude method, the sink node may communicate the new shared key to each cluster member node using its unique key, or it can do it one hop at a time using cluster keys. For the same, more complex algorithms have been developed. The authors have also offered strategies for creating shared keys among multi-hop neighbors.

A broadcast session key (BROSK) negotiation protocol has been put out by Lai et al. [62]. The master key used by BROSK is assumed to be shared by all network nodes. Node A in the WSN sends a broadcast message to its neighbor node B for initiating the creation of a shared session key. A shared session key is eventually agreed upon by the two nodes. The protocol is found to exhibit high scalability and power-efficiency.

Utilizing combinatorial design theory, Camete & Yener presented a key generation mechanism for sensor nodes in a connected network [63]. Block design approaches in combinatorics are the foundation of the key generation strategy that utilizes *combinatorial design theory* (CDTKeying). Methods such as generalized quadrangle and symmetric design are used for this purpose.

The method creates a symmetric design with the following parameters: $n^2 + n + 1$, $n + 1$, 1. It does this by using a projective plane of a finite order n, i.e., for prime powers of n. The system employs a key pool that has the size of $n^2 + n + 1$ and supports $n^2 + n + 1$ nodes. It creates $n^2 + n + 1$ key chains of size $n + 1$, each key appearing in precisely $n + 1$ number of key chains, and each pair of chains of keys sharing exactly one key. Each pair of nodes discovers precisely one key common to them after deployment. Hence, there is no chance of the existence of a shared key between any node pair. The need requirement of n being prime is a shortcoming of this claim. As a result, a given key chain size can accommodate all network sizes.

Two deterministic methods based on combinatorial design theory were suggested by Lee and Stinson: the *deterministic multiple spaces Bloms' scheme* (DMBS) and the *ID-based one-way function scheme* (IOS) [64]. In [65], they went into further detail on how combinatorial set systems may be used to create deterministic key pre-distribution methods for WSNs.

A deterministic key management approach has been proposed by Chan and Perrig [66]. The proposed scheme is based on the generation of pair-wise shared keys among the neighboring nodes in a WSN. A novel technique, *peer intermediaries for key establishment* (PIKE), is utilized for arranging all the N sensor nodes in a network in the form of a 2-D space as shown in **Figure 2**, with each node's coordinate being (x, y) where, $x, y \in \{0, 1, \ldots, \sqrt{N} - 1\}$. There are $2(\sqrt{N} - 1)$ nodes with identical x or y coordinate values while each of these nodes has distinct pair-wise keys. An intermediary node that has one of its coordinates identical to both nodes is designated dynamically as the intermediate router. The role of the router is to route the key from two nodes that do not share a common coordinate. However, the safe connectivity of

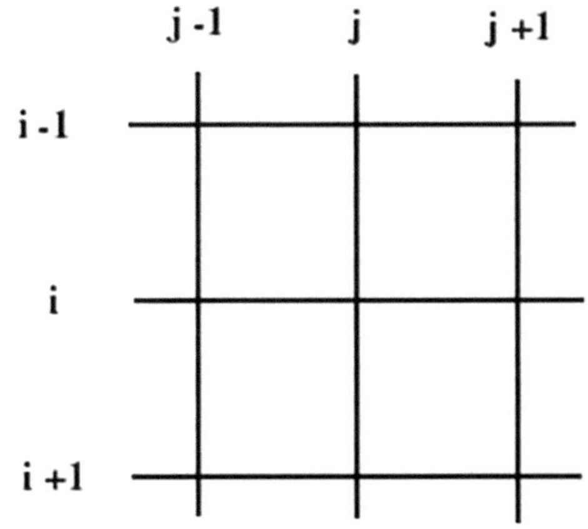

Figure 2.
The grid structure for node placement in the PIKE protocol [66].

the scheme is only $2 / \sqrt{N}$. This implies that every node should generate a key for all its neighboring nodes possibly utilizing multi-link routes. As a result, the communication overhead of the method will be significantly high.

A *hybrid authenticated key establishment* (HAKE) technique that makes use of the computational and energy differences between a sensor node and the base station in a WSN has been put forth by Huang et al. [67]. The authors contend that a single sensor node has far less computational and energy capacity than a base station. Hence, the main cryptographic computations are delegated to the central node (i.e., the base station). Lightweight symmetric-key procedures are used on the sensor side. Elliptic curve cryptography is used by the base station and sensors to authenticate each other. In the suggested technique, a public key's validity is additionally verified using certificates. The elliptic curve scheme is the foundation for the certificates. These certificates can be used to confirm the legitimacy of sensor nodes.

A t-degree $(k + 1)$-variate symmetric polynomial is used in Zhou and Fang's *scalable key agreement scheme* (SKAS), which is a deterministic key agreement methodology for generating keys in a WSN [68].

Gandino et al. [69] proposed a key management scheme for WSNs that involves the generation of a master key. The scheme is known as the *random seed distribution with transitory master key* (RSDTMK). The master key is further used in combination with a puzzle in generating the shared keys among the nodes. The shared keys are used in establishing secure communication between any pair of nodes in the network.

3.2.2 Probabilistic key distribution schemes

The majority of key distribution techniques used in WSNs are based on distributed, probabilistic systems. A *random key pre-distribution* (RKPD) approach for WSNs has been presented by Eschenauer and Gligor [70]. It is based on probabilistic key sharing between random network nodes. Key pre-distribution, shared key discovery, and path key establishment are the three stages of the mechanism. Each sensor has a key ring installed in it during the key pre-distribution step. A wide collection of P keys is randomly selected to create the k keys that make up the key ring. The base station also keeps track of the associations between the sensor identification and the key IDs on the key ring. A pair-wise key is shared by each sensor node and the base station.

Every node identifies its neighboring node with whom it has a shared key during the phase of shared key discovery. For this, the authors proposed two strategies. The basic approach involves every node broadcasting a list of plaintext key IDs from their key rings and enabling nearby nodes to determine whether those nodes have any shared key with the node. However, an attacker can use this method to track the pattern of key sharing among the nodes. The advanced approach, unlike the basic strategy, conceals key-sharing patterns between nodes from an attacker by using the challenge-response methodology. After the second phase, a path key is finally allocated in the path key setup phase for those nodes that are within the range of communication but do not have any shared key among them. The base station can instruct all nodes to revoke the keys in the king ring of a node if the node is found to be compromised. The key revocation process is identical to the key regeneration process. For authenticating the messages from the base station, the shared keys between the base station and the nodes are used. This defends against any possible attempt of a base station impersonation attack. If a node is hacked, the likelihood of an attacker successfully attacking any connection is around k/P. It only has an impact on a few sensor nodes because, $k \ll P$. This key distribution mechanism is considered to be the fundamental method among the random key distribution techniques in WSNs. There have been several other major pre-distribution strategies put forth [71–76].

Any two neighbor nodes in the basic random key management scheme must locate a single shared key from their key rings to create a safe link during the key configuration phase. However, Chan et al. found that raising the key ring's degree of key overlap might improve the network's resistance to node capture [72]. The authors' suggested a pre-distribution approach *for q-composite random keys* (QCRK). To create a safe link between any two neighbor nodes, they must share at least q common keys during the key establishment step. To improve the fundamental random key management technique, they also included a key update step. Let us say that following the key establishment step, A and B have a secure link, and the secure key is k from the key pool P. The security of the link between A and B is at risk if any of those nodes are taken over since k could be stored in the keyring memory of some other nodes in the network. The communication key between A and B should thus be updated rather than utilizing a key from the key pool. The authors have included a *multi-path key reinforcement* for the key update as a solution to this issue. If an opponent wishes to recover the communication key in this scenario, he or she must listen in on every disjoint link that connects nodes A and B. An additional layer of security is added to the system by using a *random pair-wise key management* approach for node-to-node authentication.

Typically, both nodes must broadcast their key indices or use a challenge-response mechanism to uncover common keys to determine whether the key sets of two nodes cross. Such techniques involve a significant communication overhead. By connecting a node's key indices and identification, Di Pietro et al. [74] proposed an *extended random key distribution* (ERKD) system. For instance, the key indices for each node are calculated as $g(ID, i)$ for $i = 1, 2, \ldots, N$, where ID is the node identity. Each node is given a *pseudo-random number generator*, denoted by $g(x, y)$. By confirming its node identification, other nodes can determine which key is in its key set.

Du et al. proposed a *deployment knowledge-based random key distribution* (DKRKD) system that makes use of deployment knowledge of WSNs and avoids irrelevant key assignments [75]. The authors contended that in many practical cases, some deployment knowledge may be accessible a priori which may be gainfully exploited in designing a key distribution protocol. The proposed protocol is found to significantly enhance the performance of WSNs and made the networks robust against adversarial attacks.

A polynomial-based key predistribution strategy for *group key pre-distribution* (GKPD) that may be used for WSNs was presented by Blundo et al. [76]. The bivariate t-degree polynomial presented in (1) is generated at random by the key setup server.

$$f(x, y) = \sum_{i=0}^{t} \sum_{j=0}^{t} a_{ij} x^i y^j \tag{1}$$

The bivariate polynomial is generated $\mathbf{F}_q$ over a finite field q, where q is a prime big enough to hold a key for cryptography. By selecting $a_{ij} = a_{ji}$ a symmetric polynomial, $f(x, y) = f(y, x)$, is obtained. It is expected that every sensor node has a distinct, integer-valued, non-zero identity. Each sensor node u, which has a share of polynomial $f(u, y)$ is loaded with the coefficients of the polynomial $f(u, y)$. Nodes u and v broadcast their IDs when they need to create a shared key. By computing $f(u, y)$ at $y = v$, node u may then derive $f(u, v)$, and node v can compute $f(v, u)$ by evaluating $f(v, y)$ at $y = u$. The shared key between nodes u and v has been determined to be $K_{uv} = f(u, v) = f(v, u)$ because of the polynomial symmetry. A bivariate polynomial of degree t is also $(t + 1)$-secure. To reconstruct the polynomial, an adversary must compromise at least $(t + 1)$ nodes that have the same key shares.

A polynomial *pool-based key pre-distribution* (PPKP) strategy has been put out by Liu et al. [73]. Additionally, there are three stages to the scheme: setup, direct key establishment, and path key creation. The setup server creates a set F of bivariate t-degree polynomials over the finite field $\mathbf{F}_q$ at random during the setup phase. The setup server selects a subset of polynomials $F_i \subseteq F$ for each sensor node and allocates the polynomial shares of these polynomials to node i. The sensor nodes locate a common polynomial with other sensor nodes during the direct key formation step and then create a pair-wise key using the polynomial-based key predistribution strategy described in [76]. The phase of the path key establishment is comparable to that of the fundamental random key management method. The key predistribution schemes based on random subset assignment and the grid-based key predistribution scheme are also described and analyzed in the paper. Additionally, the suggested framework enables the investigation of many instantiations.

Blom's key predistribution approach [77] is used in Du et al.'s *multiple-space key pre-distribution* (MSKP) system [71]. The system in [73] is based on a set of bivariate t-degree polynomials, whereas the scheme in [71] is based on Blom's approach. This is the main distinction between the schemes presented in [71, 73]. The suggested approach enables any pair of network nodes to locate a pair-wise secret key. The network is completely safe as long as no more than λ nodes are attacked. The base station then generates a random $(\lambda + 1) \times N$ matrix G over a finite field $GF(q)$, and an $N \times (\lambda + 1)$ matrix $A = (D. G)^T$, where $(D. G)^T$ is the transpose of the matrix $D. G$. Matrix D must be maintained a secret and must not be revealed to attackers. It is simple to confirm that $A. G$ is a symmetric matrix using (2).

$$A.G = (D.G)^T.G = G^T.D^T.G = G^T.D.G = (A.G)^T \tag{2}$$

Hence, $K_{ij} = K_{ji}$. K_{ij} (or K_{ji}) is intended to serve as the pair-wise key connecting nodes i and j. The two procedures listed below are completed in the pre-distribution phase for any sensor node k to perform the aforementioned computation: The kth column of matrix G and the kth row of matrix A are both stored at node k, respectively. The pair-wise key between nodes i and j must then be determined. To do this, nodes i and j swap their private rows of A before computing K_{ij} and K_{ji}, respectively. Each sensor node in the proposed approach is loaded with G and τ unique D matrices that are selected from a sizable pool of ω symmetric matrices $D_1, D_2, \ldots, D_\omega$ of dimension $(\lambda + 1) \times (\lambda + 1)$. The jth row of A_i should be stored at this node after computing the matrix $A_i = (D_i.G)^T$, for each D_i. Each node must determine whether it shares any space with neighbors after deployment. If the nodes discovered that they shared a space, they could use Blom's approach to create a pair-wise key. The plan is adaptable and scalable. Additionally, compared to the plan put forth in [73], it is far more durable to node capture.

A *lightweight polynomial-based key management* (LPKM) strategy for distributed WSN was put out by Fan et al. [78]. In addition to providing secure one-to-one and many-to-one communications using polynomial-based keys (such as the pairwise key, cluster key, and group key), this protocol also provided authentication using a probabilistic local broadcast authentication protocol among nearby nodes.

To provide the security of personal key shares, Wang et al. [79] presented a *hash-chain-based key management* (HCKM) strategy that was inspired by polynomials. It employs p-degree polynomial $F(x)$ to provide safe communication between and within classes. Consider a sensor network with two groups, G_1 and G_2, with the first group being G_1. If a member of group G_1 uses the key $P(v)$ to encrypt the multicast message for group G_2 members. The group controller gives a polynomial to each member of groups G_1 and G_2 so they may use it to decode this message using the key $P(x)$ that members of group G_2 obtained from members of group G_1. A revocation polynomial and a specific one-way hash function are utilized in this key distribution scheme's defense against the *collusion* attack. The one-way hash chain technique of generating the revocation polynomial is used to update the broadcast transmission. This strategy reduces communication costs and eliminates the collusion attack.

A key management system based on *polynomials by self-healing keys* (PSHK) has been presented by Sun et al. [80]. The enhanced polynomials and broadcast authentication technique can offer collision resistance and secure communication. The pairwise keys between the controller node and other sensor nodes are produced using a collection of sliding windows and enhanced polynomials. *Sch-I* and *Sch-II*, two distinct strategies, were also put forth. The *Sch-I* technique puts forward the notion that the controller node and other sensors establish and share pairwise keys. *Sch-I* may be dynamically updated in response to the network. *Sch-I* rejects the vulnerability since other nodes are unaware of this polynomial. A one-way hash function provides *forward security*, whereas a *modified polynomial* provides *backward security*. *Sch-II* enhances security by removing the hash chain. By using this approach, they were able to increase *collision resistance* while avoiding the drawbacks of acceding polynomials.

Chebyshev polynomials [81] are a novel key management strategy that Ramkumar and Singh [82] have used to create keys for the nodes. To protect message communications, the proposed scheme, known as *key management using Chebyshev polynomial* (KMCP), utilizes the features of Chebyshev's polynomials.

A novel, efficient, and *dynamic key management* (DKM) strategy for sensor networks was presented by Zhou and Yang [83]. To create effective keys, a mix of trivariate symmetric polynomials, ECC, and p-degree polynomials were used. The key is dynamically updated using a time slice approach. The communication overhead in the key distribution scheme is minimized by using a one-way hash chain among the nodes.

Jing et al. present a *fully homomorphic encryption-based key generation* (FHEKG) scheme [84]. It produced paired keys using *homomorphic encryption* [85]. The network is protected against node capture attempts using this strategy. These pairs of keys are strong, random, and unique thanks to the characteristics of an asymmetric polynomial, which satisfies the criteria of a suitable key management method.

To create paired keys among sensor nodes, Zhan et al. suggested a system using an *equation-based key distribution* (EKD) [86]. The sensor network communicated and delivered messages discreetly using these paired keys. There is only one solution to every equation in the set of equations. The generated keys are, therefore, compact, effective, and robust. To create private shared keys, linear equations' cutting points are employed. In sensor networks, these paired keys are used to defend the network from different threats. To avoid the high computational overhead involved in solving polynomial equations, this technique generates keys and implements key management in the network using linear equations with just two variables using the *exclusion basis system* (EBS). The benefit of this strategy is that, in contrast to other conventional key schemes, it offers a solid key setup, and other performance measures are unaffected.

Dinkar et al. proposed a key distribution scheme that is based on *symmetric polynomials* using a multivariate framework [87]. In this proposition, known as the *hybrid key management security scheme* (HKMSS), the keys shared between the central node (i.e., the sink node) and the cluster heads are derived using symmetric polynomials and matrices. A secure network is created using the protocol for future communications between the nodes. The matrices are regularly updated and stored at the sink node and the cluster heads. The matrices are updated whenever the shared key between a pair is changed. The key management scheme is found to be efficient even when the shared keys are frequently updated.

To ensure that the network remains connected always, the probability of a pair of nodes having a shared key should be carefully chosen and each sensor node has to store a variety of key materials. When sensor nodes have limited memory, this results in significant storage overhead. By lowering the quantity of key-related data that must be saved in each node and assuring a specific likelihood of key sharing between a pair of nodes, an improvement over the random key distribution scheme [70] is proposed by Hwang and Kim [88]. Instead of securing connections throughout the whole network, they plan to do it in the biggest subcomponent. The likelihood that two nodes share a key is decreased, but it is still high enough to link the largest network component.

The fundamental random key management technique was expanded by Hwang et al., who also put out a *cluster key grouping* (CGA) approach [89]. The authors also proposed an optimization of memory, energy requirement, and the level of security.

The essential components are evenly dispersed over the network's terrain in each of the key management systems that have been previously addressed. Because of the homogeneous distribution, the likelihood of secure connectivity—the sharing of a direct key by two neighboring nodes—is rather low. As a result, the creation of indirect keys will always include significant communication overhead. Two close sensor nodes can be preloaded with the same set of essential elements if the location of one of them is known. Secure connectivity might be enhanced in this way.

Liu & Ning proposed the *location-based key pre-distribution* (LBKP) scheme, in which a WSN is split into several cells of square shapes [90]. Every cell has a certain t-degree polynomial associated with it involving two variables. The polynomials of each sensor node's home cell and four cells that are both horizontally and vertically adjacent to it are pre-loaded onto each node. Two neighbor nodes can create a shared key between them after deployment if the two nodes possess a share of the same polynomial. As an illustration, the polynomial of cell C_{33} in **Figure 3** is likewise allocated to cells C_{32}, C_{34}, C_{23}, and C_{43}. Other cells' polynomials are allocated similarly. A node in C_{33}, therefore, shares some polynomial information with other nodes in the shaded regions.

Younies et al. proposed a key distribution scheme that utilizes the location information of the nodes in a WSN [91]. This technique generates keys using a location and an *exclusion-based system* (EBS). The produced keys are pairwise, randomized, and unique and are computed based on the locations of the nodes. The proposed scheme is also referred to as the *scalable, hierarchical, efficient, location-based, and lightweight* (SHELL) protocol. This technique provides key regeneration and enhances network security against various threats including node compromise and hijacking. All of the nodes share the burden of key management, hence reducing storage overhead and compute complexity. Additionally, the scheme avoids the overload on the base station. The location information of the nodes is used to derive the shared keys between the node pairs. The scheme is resistant to node collusion attacks. SHELL offers protection from the collusion attack. These key generation and distribution strategies allow for changes in network size, such as the addition or removal of nodes, as well as key refreshes that take node location into account.

Choi et al. presented the *location-dependent key management* (LDKM) scheme for key generation and distribution based on the location of the nodes in a WSN [92]. In the proposed scheme, Grid-based coordinates are used in this method to create network keys. Nine data coordinates and eight neighbor coordinates are utilized. These coordinated paired keys are established during the network's first and second stages. The sequence number of every packet that a node sends is also used. This approach offers protection against several internal and external dangers.

Zhu and Zhan argue that while random key predistribution is the most efficient way of managing keys in a WSN, security, and robustness of the network are two important issues that must be addressed in such approaches [93]. The authors propose

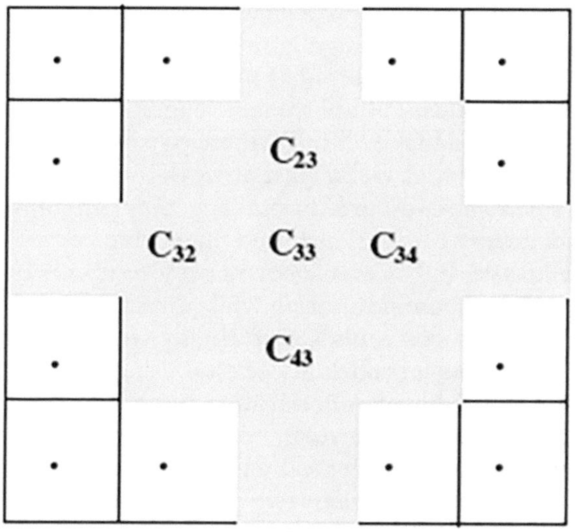

Figure 3.
The topology of a WSN divided into several cells in the LBKP scheme [90].

a *q-composite random key management* (QCRKM) approach that is based on the knowledge of the network topology.

Shi et al. propose a key management scheme in WSNs that works on dynamic authentication of the member nodes [94]. The proposed mechanism, known as the *dynamic membership authentication and key management* (DMAKM) scheme, can authenticate nodes for accessing network resources while dynamically refreshing the keys used in the authentication. The scheme preserves the forward and backward secrecy of information in the nodes and is found to be resistant to node capture attacks [95].

Cheng et al. propose a *fast multivariate polynomial-based authentication* (FMPA) and key management scheme that can combine two important functions in a WSN, (i) generation of keys, and (ii) authenticating nodes in the network based on the generated keys [96]. The authentication function has a linear complexity with the number of nodes in the network, unlike other similar schemes most of which have quadratic complexity with the network size.

Kumar and Malik present a scheme for node authentication and key distribution for WSN that supports dynamic joining and leaving of nodes [97]. The scheme proposed by the authors, known as *dynamic key management for clustered networks* (DKMCN), is suited for clustered WSNs in which the keys generated by a central node are distributed securely to the cluster member nodes via the cluster head nodes. The performance analysis of the scheme exhibited its robustness against various attacks including the node capture attacks [95].

Li et al. proposed a model of key management that consists of two layers of a key pool [98]. In the proposed scheme, known as the *one-way associated key management* (OAKM) model, the authentication of the nodes is done in two phases increasing the robustness of the key distribution and management task.

Table 1 categorizes and compares the deterministic key distribution schemes for WSNs which were discussed in this chapter. The protocols are compared for the types of keys they involve, the level of scalability and security they provide, and the processing, communication, and memory they demand. A similar comparative analysis for the probabilistic key distribution schemes is presented in **Table 2**.

In the following, some important challenges in designing efficient and secure key management schemes for WSNs are highlighted.

Memory: A key management protocol has to satisfy two goals: high security and little overhead. Several significant establishment suggestions for sensor networks have been made, however, they seldom ever fulfill these two needs. Strong security systems often demand a lot of memory, as well as fast processors and a lot of electricity. Due to the sensor platform's hardware resource limitations, they cannot readily be supported. One bit can use more energy being transmitted than being computed in a wireless context, as is widely known. In key management protocols, indirect key establishment takes place across multi-hop communication while direct key establishment just needs one-hop communication or a few rounds of it. Highly secure communication is possible when two nodes have a high probability of establishing a direct communication link with a shared key. Multi-hop communications involve more overhead and are usually less secure. However, additional key materials are needed at each node for highly secure connectivity, which is typically impracticable, especially when the network size is huge. In light of the previous two problems, memory use might be a significant barrier when developing key management procedures for a WSN. It is crucial to figure out how to lower memory use while yet keeping a certain level of security.

Prot name	Ref	Master key	Pairwise key	Path key	Cluster key	Scalability	Robustness	Proc Load	Comm load	Memory load
LKHW	[54]	Yes	Yes	No	Yes	Medium	Low	Low	Low	Low
LEAP	[60, 61]	Yes	Yes	Yes	Yes	High	Low	Low	Low	Low
BROSK	[62]	Yes	Yes	No	No	High	Low	Low	Low	Low
CDTKeying	[63]	No	Yes	No	No	High	High	Medium	Medium	Medium
IOS & DMBS	[64, 65]	No	Yes	No	No	High	High	Medium	Medium	High
PIKE	[66]	No	Yes	Yes	No	High	Low	Low	Low	High
HAKE	[67]	No	Yes	No	Yes	High	High	Medium	Medium	Medium
SKAS	[68]	No	Yes	No	No	High	High	Low	High	Low
RSDTMK	[69]	Yes	Yes	No	No	High	High	High	High	High

Table 1.
Summary of the deterministic key distribution schemes for WSNs.

Prot name	Ref	Master key	Pairwise key	Path key	Cluster key	Scalability	Robustness	Proc load	Comm load	Memory load
Basic RKPD	[70]	No	Yes	Yes	No	High	High	Medium	Medium	High
MSKP	[71]	No	Yes	No	No	High	High	Medium	Medium	High
QCRK	[72]	No	Yes	Yes	No	High	High	Medium	Medium	High
PPKP	[73]	No	Yes	No	No	High	High	Medium	Medium	High
ERKPD	[74]	No	Yes	Yes	No	High	High	High	High	High
DKRKD	[75]	No	Yes	No	No	High	High	Medium	Medium	Medium
GKPD	[76]	No	Yes	No	Yes	Low	High	Low	High	Low
LPKM	[78]	No	Yes	Yes	Yes	High	High	Medium	High	Low
HCKM	[79]	No	Yes	Yes	Yes	High	High	Medium	High	Low
PSHK	[80]	No	Yes	Yes	No	High	High	High	High	High
KMCP	[82]	No	Yes	No	No	Medium	High	High	High	High
DKM	[83]	No	Yes	Yes	No	High	High	High	Low	Medium
FHEKG	[84]	No	Yes	No	No	Low	High	High	High	High
EKD	[86]	No	Yes	No	No	Low	Medium	Low	Medium	High
HKMSS	[87]	Yes	Yes	Yes	Yes	High	High	High	High	High
CKG	[89]	No	Yes	No	No	High	High	Medium	Medium	High
LBKP	[90]	No	Yes	No	No	High	High	Medium	Medium	Medium
SHELL	[91]	No	Yes	No	Yes	High	High	High	High	High
LDKM	[92]	No	Yes	Yes	No	High	High	High	High	High
QCRKM	[93]	No	Yes	Yes	No	High	High	High	High	Medium
DMAKM	[94]	No	Yes	No	Yes	High	High	High	High	High
FMPA	[96]	No	Yes	Yes	Yes	High	High	High	High	High
DKMCN	[97]	Yes	Yes	No	Yes	Low	High	High	High	High
OAKM	[98]	No	Yes	No	Yes	High	High	High	High	High

Table 2.
Summary of the random key distribution schemes for WSNs.

3.2.2.1 End-to-end security

Symmetric key cryptography's computational efficiency is one of its main advantages. Since there can be possibly many nodes in a network, it is not a good idea for each node in a network to store a shared transport layer key with each remaining node. Hence, the majority of the existing symmetric key-based systems focus on the security of the link layer. However, many WSN applications demand secure node-to-node at the transport layer. For instance, an aggregator node may combine information from several nodes and provide the aggregated result to a designated central node (or the sink node) to minimize traffic in the network. The messages communicated between the aggregator and the central node and the source nodes and the aggregator node should both be secured and privacy-protected. But in hostile circumstances, any node is vulnerable. If one of the intermediary nodes on a route is hacked, the affected node may reveal or change the message sent down the route. End-to-end security can successfully stop hostile intermediary nodes from altering messages. Public key cryptography is more costly than symmetric key technology, but it allows end-to-end security and offers flexible management. A node equipped with both public key and symmetric key-based algorithms may use the public key algorithm to generate shared keys with other nodes in WSNs. Construction and implementation of efficient and effective public key algorithms are essential for achieving this aim so that they may be extensively applied to sensor systems. Another significant issue is how to validate the validity of public keys. Otherwise, a bad node might pretend to be any other legitimate node by stealing its public key. Identity-based encryption offers a quick solution to the issue. Pairing-based ECs are frequently employed in creating symmetric keys using the identities of the nodes since the majority of identity-based cryptographic methods now in use work on elliptic curve fields. However, the pairing procedure is extremely expensive, equal to or even more so than RSA. Therefore, the primary goals for academics are to develop quick methods and implementations.

3.2.2.2 Effective symmetric key algorithms

Because encryption and authentication based on symmetric keys are often used in the security operations of sensor nodes, there is still a need for the development of more effective symmetric key algorithms. Each packet, for instance, must be authenticated in the link layer security protocol TinySec [99], and encryption can also be activated if important packets are transferred. As a result, symmetric key algorithms that are quick and economical should be created.

3.2.2.3 Revocation and update of keys

Once a shared key is established between two nodes, it may be used as a master key to create several sub-keys for various functions (authentication and encryption). Cryptanalysis over the ciphers may eventually reveal the key if it is used over a long period. It is advisable to update keys regularly to prevent cryptanalysis of the master key and those sub-keys. But picking an update interval might be challenging. It is extremely difficult to make an educated guess as to how long it will take an adversary to disclose a key through cryptanalysis since the opponents' cryptanalysis capabilities are unknown. If the key is updated after a long interval, the associated key can also be hacked by an adversary. On the other hand, if the update interval is too short, it will involve a significant overhead of computation and communication. Key revocation is an issue that is linked. A node's key must be revoked if it turns out to be malicious. Key revocation has not, however, been properly looked into. Even though Chan et al. [100] provided a key revocation scheme that works on key pairs generated randomly using the scheme proposed in [72]. However, the proposed distributed protocol does not generalize well and hence it is difficult to use it in combination with other key distribution schemes.

3.2.2.4 Node compromise

This attack may be very detrimental for WSNs. Compromised nodes can cause extremely serious harm to WSN applications and are difficult to identify since they include all the genuine key materials. It's still unclear how to prevent node compromise. Most current security protocols make an effort to limit the impact of node breaches to a narrow region by carefully designing their protocols to minimize this impact. A hardware-based strategy, though, offers more potential. With improvements in hardware design and manufacturing methods, considerably more durable, tamper-proof, and affordable devices may be mounted on WSN. These devices may cause extremely serious damage to WSN applications and cannot be readily identified. It's still unclear how to prevent node compromise. Most current security protocols make an effort to limit the impact of node breaches to a narrow region by carefully designing their protocols to minimize this impact. A hardware-based strategy, though, offers more potential. To prevent node compromise, devices that are significantly tougher, more difficult to tamper with, and less expensive can be put on the sensor platform.

4. Conclusion

This chapter has presented a comprehensive survey of several cryptographic and key management schemes in the current literature on WSNs. Various symmetric key and public key cryptographic approaches for WSNs have been discussed and their relative merits and demerits have been highlighted. The key management protocols were categorized into three broad categories, network topology-based schemes, deterministic key distribution schemes, and probabilistic key distribution schemes. For the network topology-based schemes, the distribution or centralization of the key management task is dependent on the network architecture. In the centralized key distribution schemes, the production, distribution, and revocation of keys are all under the control of one entity under a centralized key management scheme. However, in distributed key management schemes, the tasks of key generation, distribution, and revocation are delegated to a group of nodes in the network. While the distributed key management schemes are more secure and robust, these schemes involve higher overheads of computing, communication, and storage in comparison to their centralized counterparts. In deterministic key distribution schemes, a shared key between any randomly chosen nodes is either present or absent. On the contrary, in the probabilistic key distribution schemes, the availability of a shared key between a pair of nodes is given by probability. Several key distribution schemes of both types, deterministic, and probabilistic have been discussed in this chapter. The schemes have been compared on the types of keys used and the overhead of computation, communication, and memory, the schemes involve. Some open problems for future research directions are also discussed.

Author details

Jaydip Sen
Department of Data Science, Praxis Business School, Kolkata, India

*Address all correspondence to: jaydip.sen@acm.org

References

[1] Anderson R, Kuhn M. Low cost attacks on tamper resistant devices. In: Christianson B, Crispo B, Lomas M, Roe M, editors. Security Protocols 1997, Lecture Notes in Computer Science. Vol. 1361. Berlin, Heidelberg: Springer; 1998. pp. 125-136. DOI: 10.1007/BFb0028165C

[2] Deng J, Han R, Mishra S. A performance evaluation of intrusion-tolerant routing in wireless sensor networks. In: Zhao F, Guibas L, editors. Information Processing in Sensor Networks. IPSN 2003, Lecture Notes in Computer Science. Vol. 2634. Berlin, Heidelberg: Springer; 2003. pp. 349-364. DOI: 10.1007/3-540-36978-3_23

[3] Karp B, Kung HT. GPSR: Greedy perimeter stateless routing for wireless networks. In: Proceedings of the 6th Annual International Conference on Mobile Computing and Networking; August 2000; Boston, MA, USA. Vol. 2000. New York, NY, USA: ACM Press. pp. 243-254. DOI: 10.1145/345910.345953

[4] Papadimitratos P, Haas ZJ. Secure routing for mobile ad hoc networks. In: Proceedings of the 2nd ACM Workshop on Wireless Security (WiSE'03); September 2003; San Diego, CA, USA. New York, NY, USA: ACM Press; 2003. pp. 41-50. DOI: 10.1145/941311.941318

[5] Tanachaiwiwat S, Dave P, Bhindwale R, Helmy A. Routing on trust and isolating compromised sensors in location-aware sensor networks. In: Proceedings of the 1st ACM International Conference on Embedded Networked Sensor Systems (SenSys'03), Poster Paper; November 2003' Los Angeles, CA, USA. New York, NY, USA: ACM Press; November 2003. pp. 324-325. DOI: 10.1145/958491.958542

[6] Estrin D, Govindan R, Heidemann JS, Kumar S. Next century challenges: Scalable coordination in sensor networks. In: Proceedings of the 5th Annual ACM/IEEE International Conference on Mobile Computing and Networking (MobiCom'99); August 1999; Seattle, WA, USA. New York, NY, USA: ACM Press; 1999. pp. 263-270. DOI: 10.1145/313451.313556

[7] Hu L, Evans D. Secure aggregation for wireless networks. In: Proceedings of the 2003 Symposium on Applications and the Internet Workshops; January 2003; Orlando, FL, USA. Piscataway, NJ, USA: IEEE Press; 2003. p. 384. DOI: 10.1109/SAINTW.2003.1210191

[8] Madden S, Franklin MJ, Hellerstein JM, Hong W. TAG: A tiny aggregation service for ad-hoc sensor networks. ACM SIGOPS Operating Systems Review. 2002;**36**(Special Issue): 131-146. DOI: 10.1145/844128.844142

[9] Przydatek B, Song D, Perrig A. SIA: Secure information aggregation in sensor networks. In: Proceedings of the 1st International Conference on Embedded Networked Sensor Systems (SenSys'03); November 2003; Los Angeles, CA, USA, New York. New York, NY, USA: ACM Press; 2003. pp. 255-265. DOI: 10.1145/958491.958521

[10] Shrivastava N, Buragohain C, Agrawal D, Suri S. Medians and beyond: New aggregation techniques for sensor networks. In: Proceedings of the 2nd ACM International Conference on Embedded Networked Sensor Systems (SenSys'04); November 2004; Baltimore, MD, USA. New York, NY, USA: ACM Press; 2004. pp. 239-249. DOI: 10.1145/1031495.1031524

[11] Ye F, Luo LH, Lu S, Zhang L. Statistical en-route detection and filtering of injected false data in sensor networks. IEEE Journal on Selected Areas in Communications. 2005;**23**(4): 839-850. DOI: 10.1109/JSAC.2005.843561

[12] Ganeriwal S, Balzano LK, Srivastava MB. Reputation-based framework for high integrity sensor networks. ACM Transactions on Sensor Networks. June 2008;**4**(3):1-37. DOI: 10.1145/1362542.1362546

[13] Liang Z, Shi W. Enforcing cooperative resource sharing in untrusted P2P environment. ACM Journal of Mobile Networks and Applications. 2005;**10**(6):971-983. DOI: 10.1007/s11036-005-4453-5

[14] Sen J, Roy Chowdhury P, Sengupta I. A distributed trust establishment scheme for mobile ad hoc networks. In: Proceedings of 2007 International Conference on Computing: Theory and Applications (ICCTA'07); March 2007; Kolkata, India. Piscataway, NJ, USA: IEEE Press. pp. 51-58. DOI: 10.1109/ICCTA.2007.3

[15] Sen J. A distributed trust and reputation framework for mobile ad hoc networks. In: Meghanathan N, Boumerdassi S, Chaki N, Nagamalai D, editors. Recent Trends in Network Security and Applications. CNSA 2010, Communications in Computer and Information Science. Vol. 89. Berlin, Heidelberg: Springer. pp. 538-547. DOI: 10.1007/978-3-642-14478-3_54

[16] Sen J. A trust-based detection algorithm of selfish packet dropping nodes in a peer-to-peer wireless mesh network. In: Meghanathan N, Boumerdassi S, Chaki N, Nagamalai D, editors. Recent Trends in Network Security and Applications. CNSA 2010, Communications in Computer and Information Science. Vol. 89. Berlin, Heidelberg: Springer. pp. 528-537. DOI: 10.1007/978-3-642-14478-3_53

[17] Sen J. A distributed trust management framework for detecting malicious packet dropping nodes in a mobile ad hoc network. International Journal of Network Security and its Applications (IJNSA). 2010;**2**(4):82-104. DOI: 10.5121/ijnsa.2010.2408

[18] Sen J. A distributed trust mechanism for mobile ad hoc networks. In: Proceedings of 2006 International Symposium on Ad Hoc and Ubiquitous Computing; December 2006; Mangalore, India. Piscataway, NJ, USA: IEEE Press. pp. 62-67. DOI: 10.1109/ISAHUC.2006.4290649

[19] Zhu H, Bao F, Deng RH. Computing of trust in wireless networks. In: Proceedings of IEEE 60th Vehicular Technology Conference (VTC'04-Fall); September 2004; Los Angeles, CA, USA. Piscataway, NJ, USA: IEEE Press; 2005. pp. 2621-2624. DOI: 10.1109/VETECF.2004.1400531

[20] Anderson R, Kuhn M. Tamper resistance: A cautionary note. In: Proceedings of the 2nd USENIX Workshop on Electronic Commerce; November 1996; Oakland, CA, USA. Vol. 2. New York, NY, USA: ACM Press; 1996. p. 1

[21] Anderson R, Kuhn M. Low cost attacks on tamper resistant devices. In: Proceedings of the 5th International Workshop on Security Protocols (IWSP); April 1997; Paris, France, Lecture Notes in Computer Science (LNCS). Heidelberg, Germany: Springer-Verlag; 1998. pp. 125-136

[22] de Meulenaer G, Standaert F-X. Stealthy compromise of wireless sensor

nodes with power analysis attacks. In: Chatzimisios P, Verikoukis C, Santamaría I, Laddomada M, Hoffmann O, editors. Mobile Lightweight Wireless Systems. Mobilight 2010. Lecture Notes of the Institute for Computer Sciences, Social Informatics and Telecommunications Engineering. Vol. 45. Berlin, Heidelberg: Springer; 2010. pp. 229-242. DOI: 10.1007/978-3-642-16644-0_21

[23] Sen J. Routing security issues in wireless sensor networks: Attacks and Defense. In: Tan YK, editor. Sustainable Wireless Sensor Networks. London, UK, London: IntechOpen; 2010. pp. 279-309. DOI: 10.5772/12952

[24] Akyildiz IF, Su W, Sankarasubramaniam Y, Cayirci E. A survey on sensor networks. IEEE Communications Magazine. 2002;**40**(8): 102-114. DOI: 10.1109/MCOM.2002. 1024422

[25] Sen J. Security in wireless sensor networks. In: Khan S, Pathan A-SK, Alrajeh NA, editors. Wireless Sensor Networks: Current Status and Future Trends. USA: CRC Press, Taylor & Francis Group; 2012. pp. 407-460. DOI: 10.1201/b13092-21

[26] Seshadri A, Perrig A, Van Doorn L, Khosla P. SWATT: Software-based attestation for embedded devices. In: Proceedings of IEEE Symposium on Security and Privacy; Berkeley, CA, USA. Piscataway, NJ, USA: IEEE Press; 2004. pp. 272-282. DOI: 10.1109/ SECPRI.2004.1301329

[27] Wang X, Gu W, Chellappan S, Schoseck K, Xuan D. Lifetime optimization of sensor networks under physical attacks. In: Proceedings of IEEE International Conference on Communications (ICC'05); May 2005; Seoul, South Korea. Vol. 5. Piscataway, NJ, USA: IEEE Press. pp. 3295-3301. DOI: 10.1109/ICC.2005.1495032

[28] Wang X, Chellappan S, Gu W, Yu W, Xuan D. Search-based physical attacks in sensor networks. In: Proceedings of the 14th International Conference on Computer Communications and Networks (ICCCN'05); October 2005; San Diego, CA, USA. Piscataway, NJ, USA: IEEE Press. pp. 489-496. DOI: 10.1109/ ICCCN.2005.1523922

[29] Wood AD, Stankovic JA. Denial of service in sensor networks. IEEE Computer. 2002;**35**(10):54-62. DOI: 10.1109/MC.2002.1039518

[30] Malan DJ, Welsh M, Smith MD. A public-key infrastructure for key distribution in TinyOS based on elliptic curve cryptography. In: Proceedings of the 1st IEEE Communications Society Conference on Sensor and Ad Hoc Communications and Networks (IEEE SECON'04); October 2004; Santa Clara, CA, USA. Piscataway, NJ, USA: IEEE Press; 2005. pp. 71-80. DOI: 10.1109/ SAHCN.2004.1381904

[31] Rivest RL, Shamir A, Adleman L. A method for obtaining digital signatures and public-key cryptosystems. Communications of the ACM. 1983; **26**(1):96-99. DOI: 10.1145/357980. 358017

[32] Carman DW, Krus PS, Matt BJ. Constraints and Approaches for Distributed Sensor Network Security. Technical Report 00-010. Glenwood, MD: NAI Labs, Network Associates Inc.; 2000

[33] Brown M, Cheung D, Hankerson D, Hernandez JL, Kirkup M, Menezes A. PGP in constrained wireless devices. In: Proceedings of the 9th USENIX Security

Symposium; August 2000; Denver, CO, USA. Vol. 9. New York, NY, USA: ACM Press. p. 19

[34] Gura N, Patel A, Wander A, Eberle H, Shantz S. Comparing elliptic curve cryptography and RSA on 8-bit CPUs. In: Joye M, Quisquater JJ, editors. Cryptographic Hardware and Embedded Systems—CHES 2004. CHES 2004, Lecture Notes in Computer Science. Vol. 3156. Berlin, Heidelberg: Springer; 2005. pp. 119-132. DOI: 10.1007/978-3-540-28632-5_9

[35] Gaubatz G, Kaps JP, Sunar B. Public key cryptography in sensor networks—Revisited. In: Castelluccia C, Hartenstein H, Paar C, Westhoff D, editors. Security in Ad-hoc and Sensor Networks. ESAS 2004, Lecture Notes in Computer Science. Vol. 3313. Berlin, Heidelberg: Springer; 2005. pp. 2-18. DOI: 10.1007/978-3-540-30496-8_2

[36] Wander AS, Gura N, Eberle H, Gupta V, Shantz SC. Energy analysis of public-key cryptography for wireless sensor networks. In: Proceedings of the 3rd IEEE International Conference on Pervasive Computing and Communication; March 2005; Kauai, HI, USA. Piscataway, NJ, USA: IEEE Press; 2005. pp. 324-328. DOI: 10.1109/PERCOM.2005.18

[37] Miller VS. Use of elliptic curves in cryptography. In: Williams HC, editor. Advances in Cryptology—CRYPTO '85 Proceedings. CRYPTO 1985, Lecture Notes in Computer Science. Vol. 218. Berlin, Heidelberg: Springer; 1986. pp. 417-426. DOI: 10.1007/3-540-39799-X_31

[38] Kobiltz N. Elliptic curve cryptosystems. Mathematics of Computation. 1987;**48**(177): 203-209

[39] Hoffstein J, Pipher J, Silverman JH. NTRU: A ring-based public key cryptosystem. In: Buhler JP, editor. Algorithmic Number Theory. ANTS 1998, Lecture Notes in Computer Science. Vol. 1423. Berlin, Heidelberg: Springer; 1998. pp. 267-288. DOI: 10.1007/BFb0054868

[40] Rabin MO. Digitalized Signatures and Public-Key Functions as Intractable as Factorization. Technical Report. Cambridge, MA, USA: Massachusetts Institute of Technology; 1979

[41] Elliptic Curve Cryptography, SECG Std. SEC1, 2000. Certicom Research. Available from: http://www.secg.org/collateral/sec1.pdf

[42] Kaliski B. TWIRL and RSA Key Size. Technical Note. RSA Laboratories; 2003

[43] Hankerson D, Menezes AJ, Vanstone S. Guide to Elliptic Curve Cryptography. Berlin, Heidelberg: Springer-Verlag; 2003. ISBN: 978-0-387-95273-4

[44] Freier AO, Karlton P, Kocher PC. The Secure Sockets Layer (SSL) Protocol, Version 3.0, RFC 6101. 2020. Available from: https://datatracker.ietf.org/doc/rfc6101/

[45] Hill J, Szewczyk R, Woo A, Hollar S, Culler DE, Pister K. System architecture directions for networked sensors. ACM SIGOPS Operating Systems Review. 2000;**34**(5):93-104. DOI: 10.1145/384264.379006

[46] Watro R, Kong D, Cuti S, Gardiner C, Lynn C, Kruus P. TinyPK: Securing sensor networks with public key technology. In: Proceedings of the 2nd ACM Workshop on Security of Ad Hoc and Sensor Networks (SASN'04); October 2004; Washington DC, USA. New York, NY, USA: ACM Press.

pp. 59-64. DOI: 10.1145/1029102.1029113

[47] Liu A, Ning P. TinyECC: A configurable library for elliptic curve cryptography in wireless sensor networks. In: Proceedings of 2008 International Conference on Information Processing in Sensor Networks (IPSN'08); April 2008; St. Louis, MO, USA. Piscataway, NJ, USA: IEEE Press; 2008. pp. 245-256. DOI: 10.1109/IPSN.2008.47

[48] Ganesan P, Venugopalan R, Peddabachagari P, Dean A, Mueller F, Sichitiu M. Analyzing and modeling encryption overhead for sensor network nodes. In: Proceedings of the 2nd ACM International Conference on Wireless Sensor Networks and Applications (WSNA'03); September 2003; San Diego, CA, USA, ACM Press. New York: NY, USA; 2003. pp. 151-159. DOI: 10.1145/941350.941372

[49] Menezes AJ, van Oorschot PC, Vanstone SA. Handbook of Applied Cryptography. Boca Raton, FL, USA: CRC Press; 2020. ISBN: 9780429466335. DOI: 10.1201/9780429466335

[50] Rivest RL. The RC5 encryption algorithm. In: Preneel B, editor. Fast Software Encryption. FSE 1994, Lecture Notes in Computer Science. Vol. 1008. Berlin, Heidelberg: Springer; 1995. pp. 86-96. DOI: 10.1007/3-540-60590-8_7

[51] Al-Odat ZA, Ali M, Abbas A, Khan SU. Secure has algorithms and the corresponding FPGA optimization techniques. ACM Computing Surveys. 2020;**53**(5):1-36. DOI: 10.1145/3311724

[52] Turner S, Chen L. Updated Security Considerations for the MD5 Message-Digest and the HMAC-MD5 Algorithms. RFC 6151. New York, NY, USA: ACM Press; 2011. DOI: 10.17487/RFC6151

[53] Law YW, Doumen JM, Hartel PH. Benchmarking block ciphers for wireless sensor networks. In: Proceedings of 2004 IEEE International Conference on Mobile Ad hoc and Sensor Systems; October 2004; Fort Lauderdale, FL, USA. Piscataway, NJ, USA: IEEE Press; October 2004. pp. 447-456. DOI: 10.1109/MAHSS.2004.1392185

[54] Di Pietro R, Mancini LV, Law YW, Etalle S, Havinga P. LKHW: A directed diffusion-based secure multi-cast scheme for wireless sensor networks. In: Proceedings of 2003 International Conference on Parallel Processing Workshops (ICPPW'03); October 2003; Kaohsiung, Taiwan. Piscataway, NJ, USA: IEEE Press; October 2003. pp. 397-406. DOI: 10.1109/ICPPW.2003.1240395

[55] Qin Z, Zhang X, Feng K, Zhang Q, Huang J. An efficient key management scheme based on ECC and AVL tree for large scale wireless sensor networks. International Journal of Distributed Sensor Networks. 2016;**2015**:198. DOI: 10.1155/2015/691498

[56] Foster CC. Information retrieval: Information storage and retrieval using AVL trees. In: Proceedings of 1965 ACM National Conference (ACM'65); August 1965; Cleveland, Ohio, USA. New York, USA: ACM Press; 1965. pp. 192-205. DOI: 10.1155/2015/691498

[57] Swaminathan A, Vivekanandan P. An effective lightweight key management (ELWKM) model for wireless sensor networks using distributed spanning tree structure. Asian Journal of Research in Social Sciences and Humanities. 2017;7(2): 749-770. DOI: 10.5958/2249-7315.2017.00126.5

[58] Chen HC, Christiana A. A role-based RSA key management approach in a hierarchy scheme. In: Proceedings of 2014 International Conference on Innovative Mobile and Internet Services in Ubiquitous Computing (IMIS'14); July 2014; Birmingham, UK. Piscataway, NJ, USA: IEEE Press; 2014. pp. 258-264. DOI: 10.1109/IMIS.2014.32

[59] Yao W, Han S, Li X. LKH++ based group key management scheme for wireless sensor network. Wireless Personal Communications. 2015;**83**(4): 3057-3073. DOI: 10.1007/s11277-015-2582-0

[60] Zhu S, Setia S, Jajodia S. LEAP: Efficient security mechanism for large–scale distributed sensor networks. In: Proceedings of the 10th ACM Conference on Computer and Communications Security; October 2003; Washington DC, USA. New York, NY, USA: ACM Press; 2003. pp. 62-72. DOI: 10.1145/948109.948120

[61] Zhu S, Sethia S, Jajodia S. LEAP+: Efficient security mechanisms for large-scale distributed sensor networks. ACM Transactions on Sensor Networks. November 2006;**2**(4):500-528. DOI: 10.1145/1218556.1218559

[62] Lai B, Kim S, Verbauwhede I. Scalable session key construction protocols for wireless sensor networks. In: Proceedings of IEEE Workshop on Large Scale Real-Time and Embedded Systems (LARTES'02); Austin, TX, USA; December 2002

[63] Çamtepe SA, Yener B. Combinatorial design of key distribution mechanisms for wireless sensor networks. In: Samarati P, Ryan P, Gollmann D, Molva R, editors. Computer Security – ESORICS 2004. ESORICS 2004, Lecture Notes in Computer Science. Vol. 3193. Berlin, Heidelberg: Springer; 2004. pp. 293-308. DOI: 10.1007/978-3-540-30108-0_18

[64] Lee J, Stinson DR. Deterministic key predistribution schemes for distributed sensor networks. In: Handschuh H, Hasan MA, editors. Selected Areas in Cryptography. SAC 2004, Lecture Notes in Computer Science. Vol. 3357. Berlin, Heidelberg: Springer; 2004. DOI: 10.1007/978-3-540-30564-4_21

[65] Lee, Stinson DR. A combinatorial approach to key predistribution for distributed sensor networks. In: Proceedings of 2005 IEEE Wireless Communications and Networking Conference; March 2005; New Orleans, LA, USA. Piscataway, NJ, USA: IEEE Press; 2005. DOI: 10.1109/WCNC.2005.1424679

[66] Chan H, Perrig A. PIKE: Peer intermediaries for key establishment in sensor networks. In: Proceedings of IEEE 24th Annual Joint Conference of the IEEE Computer and Communications Societies (INFOCOM'05); March 2005; Miami, FL, USA. Vol. 1. Piscataway, NJ, USA: IEEE Press. pp. 524-535. DOI: 10.1109/INFCOM.2005.1497920

[67] Huang Q, Cukier J, Kobayashi H, Liu B, Zhang J. Fast authenticated key establishment protocols for self-organizing sensor networks. In: Proceedings of the 2nd ACM International Conference on Wireless Sensor Networks and Applications (WSNA'03); September 2003; San Diego, CA, USA. New York, NY, USA: ACM Press; 2003. pp. 141-150. DOI: 10.1145/941350.941371

[68] Zhou Y, Fang Y. A scalable key agreement scheme for large scale networks. In: Proceedings of 2006 IEEE International Conference on Networking, Sensing and Control (ICNSC'06); April 23-25, 2006; Fort

Lauderdale, FL. Piscataway, NJ, USA: IEEE Press; 2006. pp. 631-636. DOI: 10.1109/ICNSC.2006.1673219

[69] Gandino F, Montrucchio B, Rebaudengo M. Key management for static wireless sensor networks with node adding. IEEE Transactions on Industrial Informatics. 2014;**10**(2): 1133-1143. DOI: 10.1109/TII.2013.2288063

[70] Eschenauer L, Gligor VD. A key-management scheme for distributed sensor networks. In: Proceedings of the 9th ACM Conference on Computer and Communications Security (CCS'02); November 2002; Washington, DC, USA. New York, NY, USA: ACM Press; 2002. pp. 41-47. DOI: 10.1145/586110.586117

[71] Du W, Deng J, Han YS, Varshney PK, Katz J, Khalili A. A pairwise key predistribution scheme for wireless sensor networks. ACM Transactions on Information and System Security. May 2005;**8**(2):228-258. DOI: 10.1145/1065545.1065548

[72] Chan H, Perrig A, Song D. Random key predistribution schemes for sensor networks. In: Proceedings of 2003 Symposium on Security and Privacy; May 2003; Berkeley, CA, USA. Piscataway, NJ, USA: IEEE Press; 2003. pp. 197-213. DOI: 10.1109/SECPRI.2003.1199337

[73] Liu D, Ning P, Li R. Establishing pairwise keys in distributed sensor networks. ACM Transactions on Information Systems Security. February 2005;**8**(1):41-77. DOI: 10.1145/1053283.1053287

[74] Di Pietro R, Mancini LV, Mei A. Random key-assignment for secure wireless sensor networks. In: Proceedings of the 1st ACM Workshop on Security of Ad hoc and Sensor Networks; October 2003; Fairfax, Virginia, USA. New York, NY, USA: ACM Press; 2003. pp. 62-71. DOI: 10.1145/986858.986868

[75] Du W, Deng J, Han YS, Chen S, Varshney PK. A key management scheme for wireless sensor networks using deployment knowledge. In: Proceedings of 2004 IEEE INFOCOM; March 2004; Hong Kong. Piscataway, NJ, USA: IEEE Press; 2004. pp. 586-597. DOI: 10.1109/INFCOM.2004.1354530

[76] Blundo C, De Santis A, Herzberg A, Kutten S, Vaccaro U, Yung M. Perfectly-secure key distribution for dynamic conferences. Information and Computation. 1998;**146**(1):1-23. DOI: 10.1006/inco.1998.2717

[77] Blom R. An optimal class of symmetric key generation systems. In: Beth T, Cot N, Ingemarsson I, editors. Advances in Cryptology. EUROCRYPT 1984, Lecture Notes in Computer Science. Vol. 209. Berlin, Heidelberg: Springer; 1985. pp. 335-338. DOI: 10.1007/3-540-39757-4_22

[78] Fan X, Gong G. LPKM: A lightweight polynomial-based key management protocol for distributed wireless sensor networks. In: Zheng J, Mitton N, Li J, Lorenz P, editors. Ad Hoc Networks. ADHOCNETS, Lecture Notes of the Institute for Computer Sciences, Social Informatics and Telecommunications Engineering. Vol. 111. Berlin, Heidelberg: Springer; 2012. pp. 180-195. DOI: 10.1007/978-3-642-36958-2_13

[79] Wang Q, Chen H, Xie L, Wang K. One-way hash chain-based self-healing group key distribution scheme with collusion resistance capability in wireless sensor networks. Ad Hoc Networks. 2013;**11**(8):2500-2511. DOI: 10.1016/j.adhoc.2013.05.015

[80] Sun X, Wu X, Huang C, Xu Z, Zhong J. Modified access polynomial based self-healing key management schemes with broadcast authentication and enhanced collusion resistance in wireless sensor networks. Ad Hoc Network. 2016;**37**:324-336. DOI: 10.1016/j.adhoc.2015.08.027

[81] Hale N. Chebyshev polynomials. In: Engquist B, editor. Encyclopedia of Applied and Computational Mathematics. Berlin, Heidelberg: Springer; 2015. pp. 203-205. DOI: 10.1007/978-3-540-70529-1_126

[82] Ramkumar KR, Singh R. Key management using Chebyshev polynomials for mobile ad hoc networks. China. Communications. 2017;**14**(11): 237-246. DOI: 10.1109/CC.2017.8233663

[83] Zhou R, Yang H. A hybrid key management scheme for heterogeneous wireless sensor networks based on ECC and trivariate symmetric polynomial. In: Proceedings of 2011 International Conference on Uncertainty Reasoning and Knowledge Engineering ; August 2011; Bali, Indonesia. New York, NY, USA: ACM Press; 2011. pp. 251-255. DOI: 10.1109/URKE.2011.6007810

[84] Jing Z, Chen M, Hongbo F. WSN key management scheme based on fully homomorphic encryption. In: Proceeding of 2017 Chinese Control and Decision Conference (CCDC'17), Chongqing, China, May 2017. Piscataway, NJ, USA: IEEE Press; 2017. pp. 7304-7309. DOI: 10.1109/ CCDC.2017.7978504

[85] Sen J. Homomorphic Encryption: Theory and Application. In: Sen J, editor. Theory and Practice of Cryptography and Network Security Protocols and Technologies. London, UK, London, UK: InTechOpen; 2013. DOI: 10.5772/56687

[86] Zhan F, Yao N, Gao Z, Tan G. A novel key generation method for wireless sensor networks based on system of equations. Journal of Network and Computer Applications. 2017;**82**: 114-127. DOI: 10.1016/j.jnca.2017.01.019

[87] Dinker AG, Sharma V. Polynomial and matrix-based key management security scheme in wireless sensor networks. Journal of Discrete Mathematical Sciences and Cryptography. 2020;**22**(8):1563-1575. DOI: 10.1080/09720529.2019.1695904

[88] Hwang J, Kim Y. Revisiting random key pre-distribution schemes for wireless sensor networks. In: Proceedings of the 2nd ACM Workshop on Security of Ad Hoc and Sensor Networks (SASN'04); October 2004; Washington, DC, USA. New York, NY, USA: ACM Press; 2004. pp. 43-52. DOI: 10.1145/1029102.1029111

[89] Hwang DD, Lai B, Verbauwhede I. Energy-memory-security tradeoffs in distributed sensor networks. In: Nikolaidis I, Barbeau M, Kranakis E, editors. Ad-Hoc, Mobile, and Wireless Networks. ADHOC-NOW, Lecture Notes in Computer Science. Vol. 3158. Berlin, Heidelberg: Springer; 2004. pp. 70-81. DOI: 10.1007/978-3-540-28634-9_6

[90] Liu D, Ning P. Location-based pair-wise key establishments for static sensor networks. In: Proceedings of the 1st ACM Workshop on Security of Ad hoc and Sensor Networks (SASN'03); October 2003; Fairfax, Virginia, USA. New York, NY, USA: ACM Press; 2003. pp. 72-82. DOI: 10.1145/986858.986869

[91] Younis MF, Ghumman K, Eltoweissy M. Location-aware combinatorial key management scheme for clustered sensor networks. IEEE Transactions on Parallel and Distributed

Systems. 2006;**17**(8):865-882. DOI: 10.1109/TPDS.2006.106

[92] Choi J, Bang J, Kim L, Ahn M, Kwon T. Location-based key management strong against insider threats in wireless sensor networks. IEEE Systems Journal. 2017;**11**(2): 494-502. DOI: 10.1109/JSYST.2015.2422736

[93] Zhu L, Zhan Z. A random key management scheme for heterogeneous wireless sensor network. In: Proceedings of 2015 International Conference on Cyber Security of Smart Cities, Industrial Control System and Communications (SSIC'15); August 2015; Shanghai, China. Piscataway, NJ, USA: IEEE Press; 2015. pp. 1-5. DOI: 10.1109/SSIC.2015.7245677

[94] Shi H, Fan M, Zhang Y, Chen M, Liao X, Hu W. An effective dynamic membership authentication and key management scheme in wireless sensor networks. In: Proceedings of 2021 IEEE Wireless Communications and Networking Conference (WCNC'21); April 2021, Nanjing, China. Piscataway, NJ, USA: IEEE Press; 2021. pp. 1-6. DOI: 10.1109/WCNC49053.2021.9417320

[95] Wang C, Wang D, Tu Y, Xu G, Wang H. Understanding node capture attacks in user authentication schemes for wireless sensor networks. IEEE Transactions on Dependable and Secure Computing. 2022;**19**(1): 507-523. DOI: 10.1109/TDSC.2020.2974220

[96] Cheng Q, Hsu C, Xia Z, Harn L. Fast multivariate-polynomial-based membership authentication and key establishment for secure group communications in WSN. IEEE Access. 2020;**8**:71833-71839. DOI: 10.1109/ACCESS.2020.2987978

[97] Kumar V, Malik N. Dynamic key management scheme for clustered sensor networks with node addition support. In: Proceedings of 2021 International Conference on Intelligent Engineering and Management (ICIEM'21); April 2021; London, UK. Piscataway, NJ, USA: IEEE Press. pp. 102-107. DOI: 10.1109/ICIEM51511.2021.9445393

[98] Li S, Zhou B, Hu Q, Wang J, Dai J, Wang W, et al. A secure scheme based on one-way associated key management model in wireless sensor networks. IEEE Internet of Things Journal. 2021;**8**(4): 2920-2021. DOI: 10.1109/JIOT.2020.3021740

[99] Karlof C, Sastry N, Wagner D. TinySec: A link layer security architecture for wireless sensor networks. In: Proceedings of the 2nd ACM International Conference on Embedded Networked Sensor Systems (SensSys'04); November 2004; Baltimore, MD. New York, NY, USA: ACM Press; 2004. pp. 162-175. DOI: 10.1145/1031495.1031515

[100] Chan H, Gligor VD, Perrig A, Muralidharan G. On the distribution and revocation of cryptographic keys in sensor networks. IEEE Transactions on Dependable and Secure Computing. 2005;**2**(3):233-247. DOI: 10.1109/TDSC.2005.37

Chapter 3

Machine Learning Algorithms from Wireless Sensor Network's Perspective

Rakesh Chandra Gangwar and Roohi Singh

Abstract

In the last few decades, wireless sensor network (WSN) emerged as an important network technology for real-time applications considering its size, cost-effectiveness and easily deployable ability. Under numerous situations, WSN may change dynamically, and therefore, it requires a depreciating dispensable redesign of the network. Machine learning (ML) algorithms can manage the dynamic nature of WSNs better than traditionally programmed WSNs. ML is the process of self-learning from the experiences and acts without human intervention or re-program. The current Chapter will cover various ML Algorithms for WSN and their pros and cons. The reasons for the selection of particular ML techniques to address an issue in WSNs, and also discuss several open issues related to 'ML for WSN'.

Keywords: wireless sensor network, machine learning, supervised learning, unsupervised learning, reinforcement learning

1. Introduction

A wireless sensor network (WSN) is a network that entails distributed tiny sensor nodes which might be supposed to screen bodily or environmental conditions and communicate with every different and alternate facts and information. Due to their small size, sensor nodes have limited computational power and energy resources. Additionally, the environment wherein they are placed varies dramatically over time. It is essentially critical to analyse sensor data as soon as it is collected. Sensor node data that has not been processed for a long duration is assumed as incomplete and inaccurate. Since WSNs are usually dynamic in nature, their topologies will frequently change. As a result of the connection loss, the network needs to add a new node. The future scope of WSN technology is bright across a wide range of application areas. In this Chapter; we list a few of the most useful ones and also how the different machine learning (ML) techniques are used in deploying the various sensor networks. However there are various other issues when it comes to these networks. The ML algorithms have been proven excellent in resolving issues particularly functional or operational issues, for example: clustering, processing of query, aggregation of data, localization, etc. While some algorithms focus on non-functional and non-operational issues like quality and efficiency of sensors, quality-of-service (QoS), security and integrity of data, etc. There are also several practical explanations that maximise resource utilisation and extend the life of the network. This chapter is prearranged as follows: Section 1 covers all the basics of WSNs along with applications issues and need of ML techniques. Section 2 presents taxonomy of machine learning algorithms and their details. Finally Section 3 concludes the chapter.

1.1 Overview

A sensor is a very small gadget that is used to capture data about a physical process or phenomenon and convert it into electrical signals that can be processed, monitored, and analysed further to get the purposeful information. Any type of information from the real environment, including temperature, pressure, light, sound, motion, position, flow, humidity, and radiation, could be referred to as a physical process.

In order to record, observe, and respond to an event or a phenomenon, a structure made up of sensors, processing units, and communication components are known as a sensor network. The controlling or observing body may be a consumer application, a government agency, a civil organisation, a military force, or an industrial entity, and the event may be connected to anything, including the physical world, an industrial environment, a biological system, or an IT (information technology) framework. Such sensor networks can be used for data collecting, surveillance, monitoring, medical telemetry, and remote sensing, etc. Sensors with a controller (base station), and a communication system make up a typical sensor network [1]. **Figure 1** illustrates what is known as a Wireless Sensor Network (WSN) when the connectivity in a sensor network is established utilising a wireless protocol.

1.2 Elements of WSN

There are two basic elements of WSN:

a. **Sensor node**: A Wireless Sensor Network (WSN) [2] consists of sensor nodes that are deployed in close proximity and frequently on massive scales, and it supports sensing, statistical processing, embedded computing and connectivity. WSNs are innately aid constraints and are chargeable for self-organising the

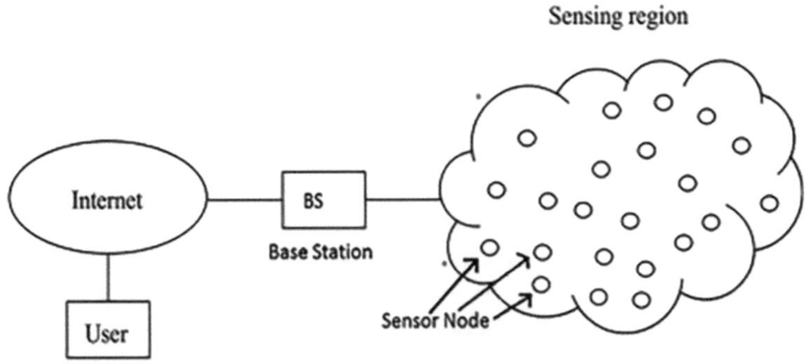

Figure 1.
A basic working model of wireless sensor network.

suitable community infrastructure frequently with multi-hop communication with them. It consists of four basic components:

- Power supply;
- Sensor;
- Processing unit; and
- Communication system.

The sensor node collects the analogue statistics from the consumer, and analogue-to-digital convertor (ADC) converts the records into the digital shape. The processing unit is the primary unit, it consists of a storage unit and a microcontroller/microprocessor, which primarily do the information processing and manipulation. It also consists of various other features like network analysis, data correlation and fusion of data from another sensor with its own. The communication system consists of any kind of system that is typically a short-range radio for data transmission and reception (**Figure 2**) [1].

a. **Network architecture**: When a massive number of sensor nodes are deployed in a large region to cooperatively screen the physical surroundings, the network of these sensor nodes is equally important. A sensor node in a WSN no longer only communicates with other sensor nodes, however, also with a Base Station (BS) using wireless communication. The base station sends instructions to the sensor nodes, and the sensor nodes perform the task by collaborating with each other. After gathering all the necessary information, the sensor nodes send the

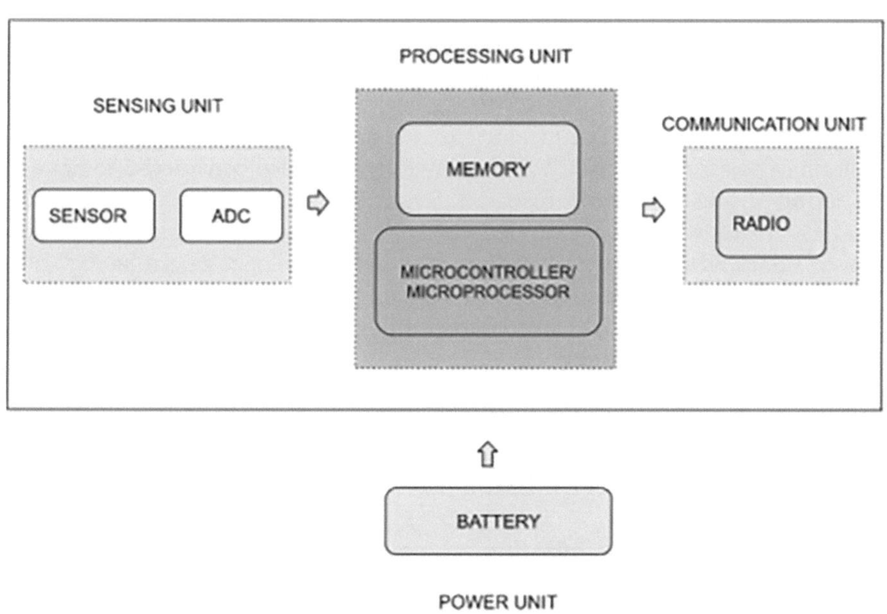

Figure 2.
Structure of a sensor node.

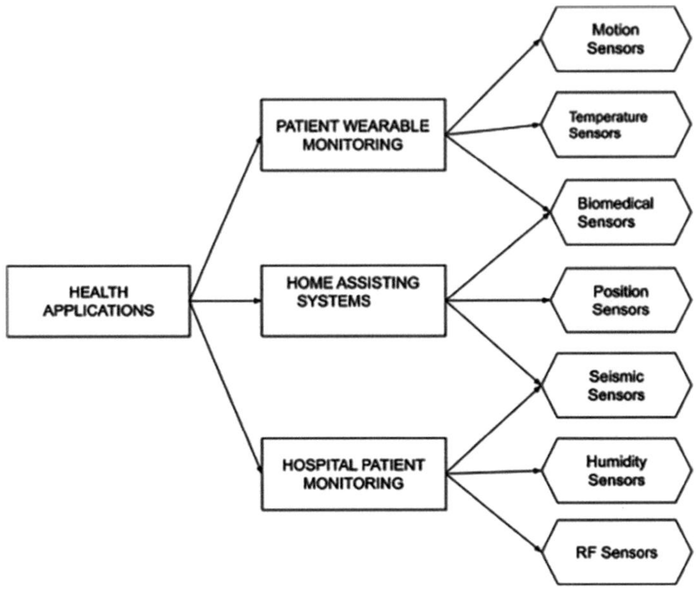

Figure 3.
Subcategories of Military applications of WSN.

data back to the base station. After receiving the records from the sensor nodes, a base station performs processing of records and sends the updated data to the person using internet.

1.3 Applications of WSN

Numerous applications [3] of WSNs are currently either in use or in the process of their development. A few applications of WSNs are listed below:

1. **Military applications**: The military area is not always the simplest the first field of human pastime that used WSNs, however, it is also considered to have the initiation of sensor network research. Smart dust is an example of those initial studies, efforts which were carried out within the late 19 if you want to broaden sensor nodes which notwithstanding their very small size might be capable of engaging in spying sports. The main subcategories of the military applications [4] of WSNs are battlefield surveillance [5], combat monitoring and intruder detection as shown in **Figure 3**.

2. **Health applications**: In the health realm, WSNs make use of superior medical sensors to display the real-time tracking of patient's vitals [6]. **Figure 4** shows the primary sub-categories of the health package of WSN.

3. **Environmental applications**: Environmental programs that seek for the monitoring of the ambient situations at adverse and far flung areas may advance with the usage of WSNs. The main sub-categories of environmental applications of WSNs such as water tracking [7], air tracking [8], and emergency alerting [9], are depicted alongside the types of sensors in **Figure 5**.

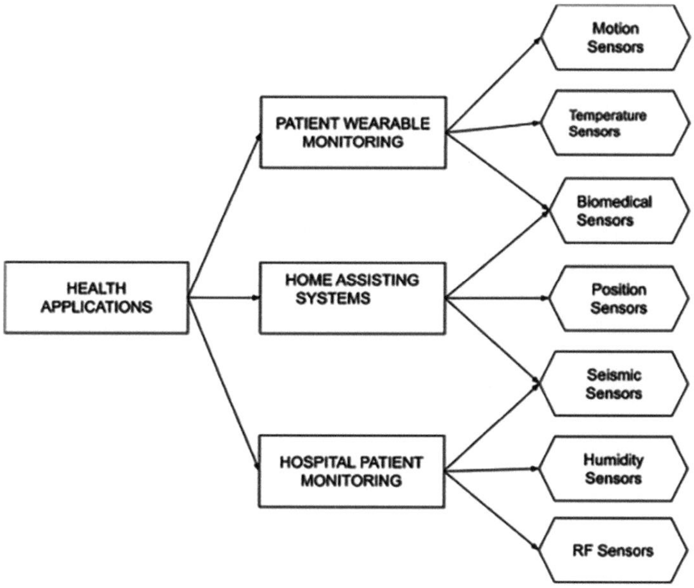

Figure 4.
Subcategories of Health Applications of WSN and the types of sensors used.

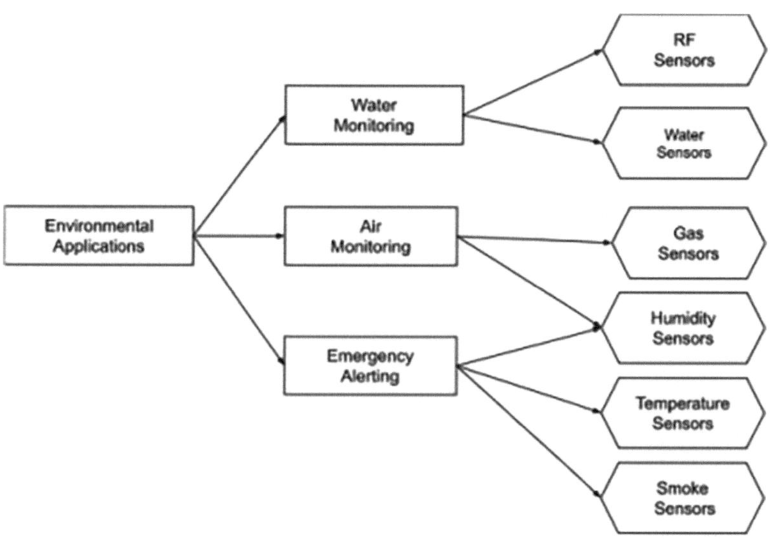

Figure 5.
Environmental Applications of WSNs and the types of sensors used by them.

4. **Flora and fauna applications**: Both flora and fauna domains are vital for every country. The animal behaviour can be studied in some crucial areas, their tracking, and to control the use of wildlife passage by the local fauna [10]. The subcategories are mentioned in **Figure 6**.

5. **Industrial applications**: The main advantage of WSNs is the absence of any wiring due to which this can be integrated up to a larger scale as well. The main sub categories of industrial applications along with their sensors are given in **Figure 7**.

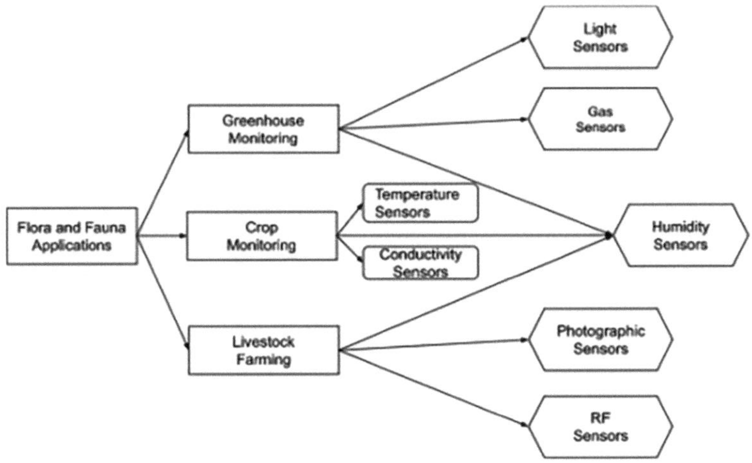

Figure 6.
Flora and Fauna Applications of WSNs and the types of sensors used by them.

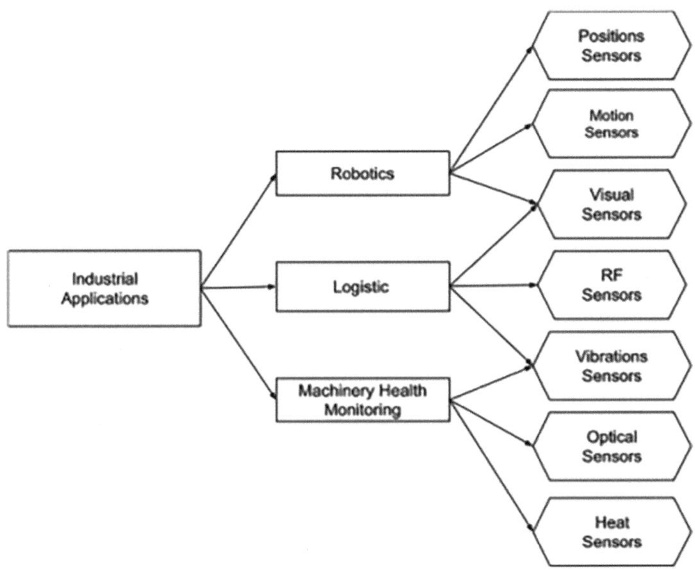

Figure 7.
Industrial Applications of WSNs and the types of sensors used by them.

6. **Urban applications**: WSNs can be used to solve the various urban problems, for example, coordination of the specialised vehicles like ambulance, fire tenders, rescue vehicles, police automobiles, logistics of public transportation, traffic management, monitoring chemical/physical environmental parameters, building security and many others (**Figure 8**) [7].

1.4 Issues in WSN

There are benefits and drawbacks to every technology. Similarly wireless sensor networks [11] also have some issues although being an excellent tool for application in many areas. The main problematic areas are: Design and the Topology.

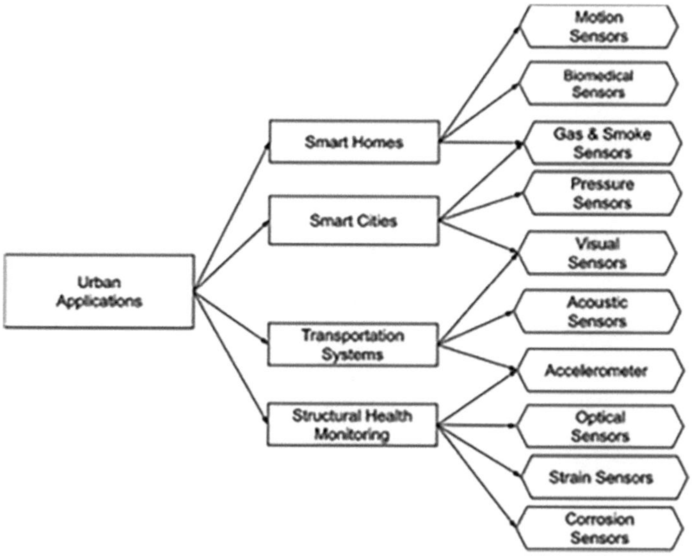

Figure 8.
Urban Applications of WSNs and the types of sensors used by them.

A. **Design issues**

1. **Fault tolerance communication**: Due to the deployment of the sensor nodes in an out of control or harsh environment, it is not always uncommon for the sensor nodes to grow to be faulty and unreliable;

2. **Scalability**: A system whose overall performance improves after adding hardware proportional to the capability added is stated to be a scalable device. The variety of sensor nodes deployed within the sensing area can be in order of hundreds or thousands or even in millions;

3. **Low latency**: Low latency is imperative as the customers expect to interact with technology in real-time with no delays. Issues with high latency and time delays can cause users to quit and go for alternatives;

4. **Transmission media**: Long range transmission is generally point-to-point and calls for high transmission power with the chance of being eavesdropped, hence for better performance and security short range transmission can be opted;

5. **Coverage problems**: Coverage problems mean how to monitor a network problem effectively. It also reflects the quality-of-service (QoS) provided by the network.

B. **Topology issues**

1. **Geographic routing**: It is one of the most extensively used technique, however, the recent studies has shown that geographic routing can sometimes be useless in actual time deployments where location estimation system introduce regional errors;

2. **Sensor holes**: A routing hole is a region in the sensor network wherein nodes are not either available or if available these cannot participate in the actual routing of the data due to various possible reasons. The mission of identifying holes is specifically difficult considering the fact that ordinary wireless sensor networks encompass lightweight, low capability nodes, which can be ignorant of their geographic place;

3. **Coverage topology**: Coverage of the sensor network represents how well the sensors monitor a field of interest where they are deployed. It is the performance measure of the network sensing capability. Connectivity represents how well the nodes communicate.

C. **Other issues are**:

1. Synchronisation;
2. Computation and energy constraints;
3. Security;
4. Cost effective;
5. Limited bandwidth/framework;
6. Node costs;
7. Power management;

1.5 Need of machine learning in WSN

Machine learning [12] is a branch of Artificial Intelligence (AI). It is basically defined as the capability of a machine to mimic the behaviour of the human being that focuses on interpreting and analysing the patterns.

Meanwhile, its focus evolved and shifted more to the algorithms, which are more achievable and reliable. The machine learning techniques have been used extensively for variety of responsibilities which includes classification, regression, biometrics (such as speech recognition, eye detection, and fingerprint detection), fraud detection, etc. The sensor nodes in the WSNs might be heterogeneous, which are designed using numerous types of sensors according to the requirements of the network. The creators of network are more vulnerable to the issues regarding aggression or collection of statistics, reliability, clustering of nodes, safety and fraud detection [13]. Wireless sensor networks keep an eye on situations that are always changing. Either external factors or the system designers themselves started this dynamic behaviour. Sensor networks frequently use machine learning techniques to adapt to such circumstances in order to avoid needless redesign. A lot of doable solutions that maximise resource usage and increase network longevity are also inspired by machine learning. Lately, the use of machine learning algorithms has been experienced in WSNs. It enhances the performance of the network without the need for reprogramming. The algorithms also extract different levels of abstractions needed to perform variety of tasks with limited or no intervention. Some of the algorithms of ML deal with the design and functional issues of the network which were stated above.

The machine learning techniques [14] are used for the following reasons:

- WSN reveal dynamic environments that change unexpectedly with time which are due to outside factors or by the designers themselves. The networks in such situations adopt the ML techniques to take away the need of unnecessary redesign;

- The machine learning additionally encourages many realistic solutions that maximise aid usage and prolong the lifespan of the network;

- The ML algorithms used to discover important correlations in the sensor data;

- WSN encouraged by means of machine studying strategies can offer low intricacy approximations for the gadget fashions, enabling its implementation within sensor nodes;

- It also offers low complexity mathematical models for complex environment;

- These are also efficiently followed for predicting future events based on the previous sensor network records.

2. Machine learning techniques for wireless sensor networks

Most of the machine learning algorithms falls into three broad categories:

- Supervised learning

- Unsupervised learning

- Reinforcement learning

2.1 Supervised learning

Supervised Learning as the name indicates it means "having a mentor" to supervise, which means that the output is already present and the input gives the output accordingly. It contains a model that is used to predict the outcomes with the help of labelled dataset. The learned relationship between the input, output and the parameters of the system is learned by system model. The training is given to this model and once it is complete the model is tested on the basis of test data and then it predicts the output with the help of labelled dataset (it is a dataset wherein the target answer is already known) (**Figure 9**) [12].

It also finds the mapping function to map the input variable and the output variable. This type of approach is used to solve diverse issues for WSN such as object targeting and localization, processing of query and event detection, medium access control, intrusion detection and security, image classification, spam filtering, data integration and security.

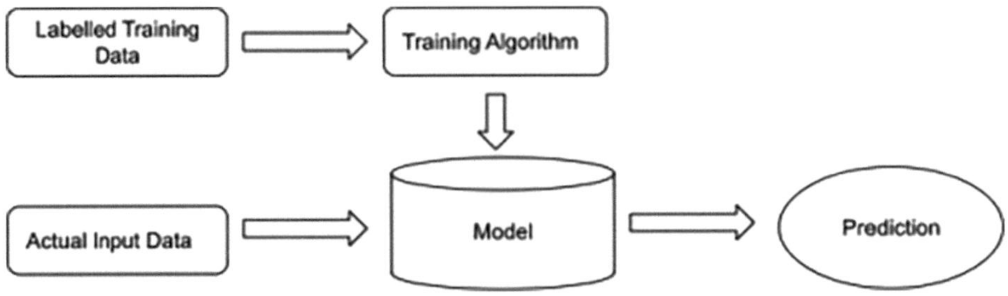

Figure 9.
Supervised learning model.

Guided learning involves the following steps:

- Initially, choose the training dataset type;
- Obtain the labelled training data;
- The data should be divided into a training dataset, a test dataset, and a validation dataset. The training dataset's input features should be identified, and they should contain sufficient information to allow the model to correctly predict the output;
- Choose an appropriate method for the model, such as a decision tree, support vector machine, etc.;
- Run the algorithm on the training dataset. Validation sets, a subset of training datasets, are occasionally needed as control parameters;
- By supplying the test set, you may determine whether the model is accurate. If the model accurately predicts the desired outcome then it is accurate.

The supervised learning can be further divided into two categories: *Regression* and *Classification.*

2.1.1 Regression algorithms

If there is a link between the input and output variables and the output has a real or continuous value, regression methods are used. This kind of learning strategy is employed in situations where a relationship between two variables and the changes in one affects the others. In simple words, "Regression exhibits a line or curve that traverses through all the data points on target-predictor graph in such a way that the distance between the data points and the regression line is small". Prediction, forecasting, time series modelling, and establishing the causal connection between variables are its key applications.

Some of the few examples of regression are as follows:

- Prediction of rain using temperature and other factors;
- Determining Market trends;
- Prediction of road accidents due to rash driving.

2.1.2 Classification algorithms

These algorithms are used when we have to classify something into group after getting output into a category. It is a method for classifying observations into several groups according to a condition. Binary classification, such as Yes-No, Male-Female, True-false, etc., refers to an algorithm's attempt to categorise data into two separate groups. Multiclass classification is the practise of choosing from more than two categories. In the classification algorithm, an input variable is transferred to a discrete output function (y) (x).

$$y = \text{categorical output}, f(x) = y$$

The primary goal of classification algorithms is to determine the category of a given dataset, and these algorithms are primarily employed to forecast the results for categorical data. Email Spam Detector is the best illustration of an ML classification method. It can also be used to categorise various objects, such as fruits, according to their taste, colour, size, etc. When given new data, a machine that has been properly trained using a classification method may predict the class with ease. It can classify everything, including fruit, automobiles, houses, signs, and more. However, there are numerous ways to perform a same task in order to predict whether a given person is male or a female, a machine has to be trained first and further there are numerous ways in doing so. For predictive analytics some of the commonly used algorithms are:

1. Decision tree;

2. Random forest tree;

3. Bayesian statistics;

4. Support vector machine (SVMs);

5. KNN.

2.1.2.1 Decision tress

Decision Tree is a supervised learning technique that is used to effectively handle non-linear data sets [15]. It is a classification learning algorithm. It is a visual representation of every scenario that could lead to a choice. It solves a problem by using a tree representation. The decision tree, as shown in **Figure 10**, resembles a tree that has two different types of nodes: the Choice (Decision) node and the Leaf node. Decisions are made using choice nodes, which have numerous branches, whereas leaf nodes are the results of those decisions and do not have any additional branches. CART, or Classification and Regression Tree Algorithm, is employed in this case to construct a decision tree. It simply asks a question and on the basis of that (either YES/NO) the tree is split further.

The fundamental idea of the decision tree is to test the most significant attribute first. The most important attribute is the one that impacts the classification of an example the most. This will ensure that we obtain the accurate classification with smaller number of tests wherein all the paths in the tree are short and the tree in general is a small one. Decision tree algorithms are used to solve many unresolved

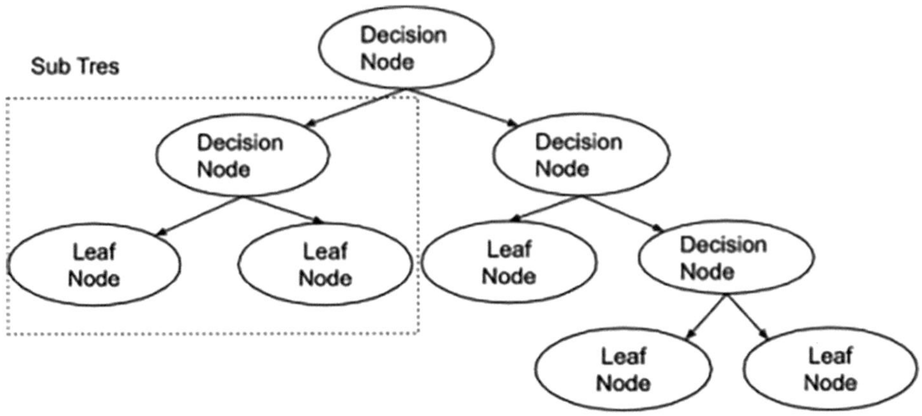

Figure 10.
Decision tree.

issues in WSNs layouts like link loss, reliability, restore, corruption rate, and mean-time-to-failure (MTTF).

2.1.2.2 How decision tree algorithm works?

In the decision tree (DT), for predicting the class of the given dataset, the algorithm begins from the foundation node i.e., the decision node or root node of the tree. The algorithm compares the value of root characteristics with the record (real dataset) characteristics and based on the comparison the node is split on. For the subsequent node, the algorithm once more compares the attribute value with the alternative sub-node and circulates similarly. It keeps the manner until it reaches the leaf node of the tree. The whole technique can be better understood by the use of below algorithm:

Step 1: Start the tree with the foundation node (let suppose S), which contains the complete dataset;
Step 2: Find the high quality attribute within the dataset using the characteristics attribute selection measure (ASM);
Step 3: Divide the foundation node(S) into the sub-nodes that contains the feasible values for the high quality attributes;
Step 4: Generate the decision tree node, which incorporates the most appropriate value or attribute;
Step 5: Recursively make new choices by using the subsets created in the step 3;

Maintain this process until a degree is reached wherein you cannot further classify the nodes known as the final node known as the leaf node.

2.1.2.3 Attribute selection measures (ASM)

Whilst enforcing a decision tree, the principle issue arises that how to select the first rate attribute for the foundation node and for sub nodes. So, to resolve such issues there may be a technique, which is known as characteristic choice measure or ASM. With this one can effortlessly choose the first-rate attribute for the nodes of the tree.

The two basic techniques for ASM are:

1. Information gain;

2. Gini Index.

Information advantage: Information gain is the clever idea for defining impurity. Impurity measure is a heuristic for selection of the splitting criterion that separates a given dataset of class labelled training tuples into individual classes. If we divide D into smaller partitions as per the outcomes of the splitting criterion, each partition should ideally be pure with all the tuples falling into each partition belonging to the same class. It is the change in entropy after the segmentation of a dataset primarily based on characteristics. It calculates how tons of information is provided about a category. Consistent with the cost of records gain, the node can be split up and build the selection tree.

The information gain can be calculated by using the following notations:

$$\mathrm{IG} = \mathrm{Entropy}\ (\mathrm{S}) - \left[\mathrm{weighted\ average}^{*}\ \mathrm{Entropy}\ (\mathrm{each\ feature})\right]$$

Entropy is the measure of randomness of the data. It can be calculated as:

$$\mathrm{E} = -(\mathrm{Probability\ of\ yes}) - (\mathrm{Probability\ of\ no})$$
$$\mathrm{E} = -\mathrm{P}\ (\mathrm{yes}) \log_2 \mathrm{P}(\mathrm{yes}) - \mathrm{P}(\mathrm{no}) \log_2 \mathrm{P}(\mathrm{no})$$

Gini Index: It is the degree of impurity or purity used at the same time as developing a decision tree for CART algorithm. An attribute with a low Gini index must be desired in comparison to the high Gini index. It only creates the binary splits, which are done by the algorithm by using the Gini index. It can be calculated by the usage of following method:

$$\mathrm{GI} = 1 - \sum_{j} \mathrm{P}_j^2$$

Some pros of using a decision tree:

1. It is straightforward to apprehend because it follows the same system, which humans observe at the same time as making any choice-related issues;

2. In comparison to different algorithms selection criteria requires much less effort for the information training throughout pre-processing;

3. A decision tree does not require any kind of normalisation of the records;

4. A decision tree does not require scaling of the statistics;

5. The missing values inside the information (datasets) also does not affect the process of building a decision tree to any extensive extent;

6. This type of tree version is very intuitive and easy to explain to the technical groups also.

Cons of using a decision tree:

1. A small exchange inside the records can reason a large trade within the shape of the choice tree inflicting instability;

2. They are vulnerable to overfitting;

3. For a decision tree some random calculation can go far more complicated compared to other algorithms;

4. This often entrails higher time to educate the model;

5. It is cost effective sometimes because of its complexity as it may contain lots of layers;

6. It is insufficient for applying regression and predicting continuous values.

Decision trees are frequently used in WSNs because they describe relationships between attributes and classes in a clear and understandable way. A decision tree's paths are made up of a series of conditions that each describes a class. Such decision tree paths can be used to develop rules that can be employed in a WSN to distinguish between different outcomes or phenomena based on measurements taken from sensed data. A decision tree's clarity provides important insights because of the open model learning approach it employs. They are also used because of their simplicity and interpretability in their regulations that can be easily derived from the shape of the tree. It also identifies hyperlink reliability, which is very successful but this set of rules works best with linearly separable records only.

2.1.2.4 Random forest tree

Random Forest tree, as shown in **Figure 11**, is a famous gadget mastering algorithm that belongs to the supervised learning approach. It is able to be used for both type and Regression issues in ML. it is primarily based on the concept of ensemble

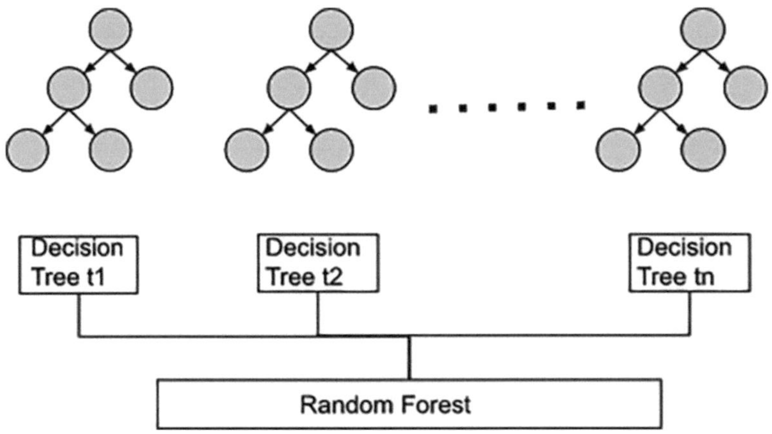

Figure 11.
Random forest tree.

gaining knowledge, which is a system of mixing a couple of classifiers to solve a complicated trouble and to enhance the performance of the model.

Random forest is a classifier that incorporates a number of decision bushes on numerous subsets of the given dataset and takes the common to enhance the predictive accuracy of that dataset. In place of relying on one selection tree, this algorithm takes the prediction from every tree and based on most people votes of predictions, and it predicts the final output.

2.1.2.5 How does random forest algorithm work?

Random forest algorithm works in two-section first is to create the random forest by using combining N choice tree, and 2-D is to make predictions for every tree created in the first section.

The running method may be defined as follows:

Step-1: Pick out random k records points from the education set;
Step-2: Construct the decision timber associated with the selected information factors (Subsets);
Step-3: Choose the variety N for choice bushes that you want to construct;
Step-4: Repeat Step 1 and 2;
Step-5: For brand new statistics points, discover the predictions of each selection tree, and assign the brand new data factors to the class that wins the majority votes.

Some pros of using random forest algorithm:

1. Sturdy to outliers;

2. Works well with non-linear records;

3. Decrease threat of overfitting;

4. Runs effectively on a huge dataset;

5. Higher accuracy than other type algorithms;

6. No function scaling required;

7. This algorithm can robotically take care of lacking values;

8. This algorithm may be very stable. Despite the fact that a brand new data factor is delivered in the dataset, the general set of rules is not always affected a great deal since the new facts might also impact one tree, but it is very difficult for it to affect all the timber.

9. It is much less impacted by way of noise.

Some cons of using random forest algorithm:

1. Random forests are found to be biased while handling express variables;

2. Sluggish training;

3. Now not suitable for linear techniques with a whole lot of sparse functions.

WSN is a difficult problem due to the diversity of deployment and the restrictions within the sensors' sources. This supervised device mastering-based total approach is considered to scrutinise the behaviour of sensors through their statistics for the detection and prognosis of faults. Maximum of the faults that generally arise in WSN are considered: handover, glide, spike, erratic, information-loss, stuck, and random fault. A hybrid strategy was put forth [16] for real-time network intrusion detection systems (NIDS). For feature selection, they use the random forest (RF) algorithm. In order to remove the unnecessary features, RF presents the variable importance as numeric values. The experimental findings demonstrate that the new strategy is quicker and lighter than the prior methods while still ensuring high detection rates, making it appropriate for real-time NIDS.

2.1.2.6 Bayesian statistics

Bayesian records are the mathematical method for calculating possibilities wherein inferences are subjective and get updated while extra facts are delivered. This record is in comparison with classical or frequentist information where probability is computed through comparing the frequency of a specific random occasion for an extended length of repeated trials where inferences are intended to be the goal. Those statistical inferences are the manner of extracting conclusions out of massive datasets via studying a small portion of sample statistics. For this, the data professionals:

- First examine the pattern information and extract the belief, this is called as prior inference;
- After this, they check another sample of records and revise their end, this revised data is known as posterior inferences.

As Bayesians, a concept of a notion, called a previous, gain some information and use it to update the notion. The final results are called a posterior. As attain even greater facts, the antique posterior becomes a brand new prior and the cycle repeats.
This system employs the Bayes rule:

$$P(A|B) = P(B|A)^{*}\ P(A)/P(B)$$

P(A|B), examine as "possibility of A given B", shows a conditional chance: how possibly is A if B happens.
In WSNs, these styles of Bayesian learners are useful for assessing event consistency. Numerous variations of Bayesian newcomers allow better getting to know of relationships, consisting of Gaussian combination fashions, Dynamic Bayesian Networks, Conditional Random Fields as well as Hidden Markov fashions.

2.1.2.7 Support vector machines (SVMs)

One of the most well-liked supervised learning algorithms, Support Vector Machine (SVM) [17] is used to solve Classification and Regression problems.

However, it is largely employed in Machine Learning Classification issues. The SVM algorithm's objective is to establish the best line or decision boundary that can divide n-dimensional space into classes, allow to quickly classifying fresh data points in the future. A hyper plane is the name given to this optimal decision boundary.

SVM selects the extreme vectors and points that aid in the creation of the hyper plane. Support vectors, which are used to represent these extreme instances, form the basis for the SVM method. Take a look at the **Figures 12** and **13**, where two distinct categories are identified using decision boundary:

In n-dimensional space, there may be several lines or decision boundaries used to divide classes; however, the optimal decision boundary for classifying the data points must be identified. The hyperplane of SVM is a name for this optimal boundary. The features of dataset determine the dimensions of hyper plane, therefore, if there are just two features, as shown in **Figure 12** [17], the hyperplane will be a straight line. Additionally, if there are three features, the hyperplane will only have two dimensions.

Support Vectors: The data points or vectors that are closer to the hyperplane and influence the position and orientation of the hyperplane. The SVM method aids in identifying the ideal decision boundary or region, often known as a hyperplane. The SVM algorithm determines which line from each class is closest to the other. Support

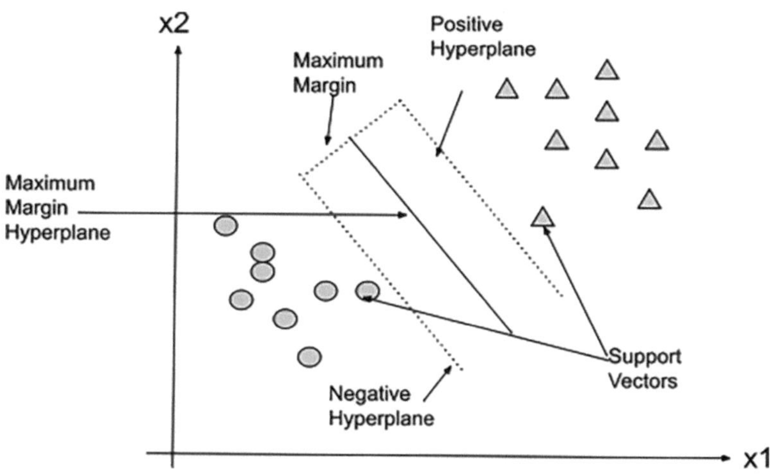

Figure 12.
Support Vector Machine (SVM).

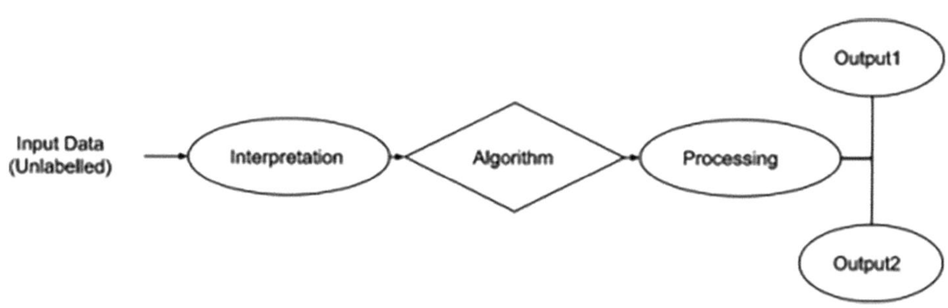

Figure 13.
Unsupervised learning algorithm.

vectors are the names for these points. Margin is the distance between the hyperplane and the vectors. Maximising this margin is the aim of SVM. The ideal hyperplane is the one with the largest margin.

SVM can be categorised in two different ways:

Linear SVM: Linear SVM is used for linearly separable records, which means that if a dataset may be categorised into instructions by the usage of a single straight line, then such facts is called as linearly separable statistics, and classifier is used called as Linear SVM classifier.

Non-linear SVM: Non-Linear SVM is used for non-linearly separated information, which means that if a dataset cannot be categorised by way of the use of a directly line, then such data is termed as non-linear statistics and classifier used is referred to as Non-linear SVM classifier.

Face identification, image classification, text categorisation, grouping of portrayals, bio-informatics, handwriting remembrance, etc. may all be done using the SVM method.

Some pros of using the SVM algorithm:

1. Support vector machines perform similarly well when there is a discernible margin of class dissociation;

2. High dimensional rooms are more productive;

3. When the quantity of dimensions exceeds the quantity of examples, it works well;

4. This algorithm is very much similar to memory systematic.

Some cons of using this algorithm:

1. For huge data sets, the support vector machine approach is unacceptable;

2. When the target classes overlap and the data set has more sound, it does not operate very well;

3. The support vector machine will perform poorly when there are more attributes for each data point than there are training data specimens;

4. There is no classification error since the support vector classifier places data points above and below the classifying hyperplane.

2.1.2.8 K-Nearest Neighbour (K-NN)

K-Nearest Neighbour is one of the most effective machines gaining knowledge of algorithms based totally on Supervised getting to know method. The K-NN algorithm makes the assumption that the new case and the existing cases are comparable, and it places the new instance in the category that is most like the existing categories. A new data point is classified using the K-NN algorithm based on similarity after all the existing data has been stored. This means that utilising the K-NN method, fresh data can be quickly and accurately sorted into a suitable category. Although the K-NN approach is most frequently employed for classification problems, it can also be utilised for regression. Since K-NN is a non-parametric technique, it makes no assumptions about the underlying data [18].

It is also known as a lazy learner algorithm since it saves the training dataset rather than learning from it immediately. Instead, it uses the dataset to perform an action when classifying data. The KNN method simply saves the information during the training phase, and when it receives new data, it categorises it into a category that is quite similar to the new data.

The K-NN algorithm is excellent for WSN query processing jobs because of its simplicity.

The following algorithm can be used to describe how the K-NN works:

Step 1: Decide on the neighbours K-numbers;
Step 2: Calculate the Euclidean distance (or Hamming Distance) between K neighbours. The distance between two points, which we have already examined in geometry, is known as the Euclidean distance;
E.g.: Let there be two points $A(x_1,y_1)$ and $B(x_2,y_2)$. Now the Euclidean distance between them can be calculated as: $ED = \sqrt{\left[(x_2 - x_1)^2 + (y_2 - y_1)^2\right]}$.
Step 3: Based on the determined Euclidean distance, select the K closest neighbours;
Step 4: Count the number of data points in each category among these k neighbours;
Step 5: Assign the fresh data points to the category where the neighbour count is highest;
Step 6: Model is complete.

Some pros of using the KNN algorithm:

1. It is straightforward to put in force;

2. It's far strong to the noisy schooling records;

3. It can be extra powerful if the training facts are huge.

Some cons of using the KNN algorithm:

1. K's value must always be determined, and sometimes that can be difficult;

2. The high computation cost is caused by the need to determine the separation between each data point for each training sample.

2.2 Unsupervised learning

In supervised machine learning, models are trained on labelled data while being watched over by training data. However, there may be several instances when lacking labelled data and need to identify hidden patterns in the supplied dataset. Therefore, one needs unsupervised learning strategies to handle these kinds of problems in machine learning. Unsupervised learning is a subcategory of machine learning wherein models are trained using unlabelled datasets and are free to operate on the data without being checked by a human observer.

Because unlike supervised learning, one has the input data but no corresponding output data, unsupervised learning cannot be used to solve a regression or classification problem directly. Finding the underlying structure of a dataset, classifying the data based on similarities, and representing the dataset in an unsupervised way and in the compressed format are the objectives of unsupervised learning [19].

The following are a few key arguments for the significance of unsupervised learning:

1. Finding valuable insights from the data is made easier with the aid of unsupervised learning;

2. Unsupervised learning is considerably more like how humans learn to think via their own experiences, which brings it closer to actual artificial intelligence;

3. Unsupervised learning is more significant because it operates on unlabelled and uncategorized data;

4. Unsupervised learning is necessary to handle situations when the input and output are not always the same in the real world;

According to the **Figure 13**, input data is unlabelled, which means that neither its category nor any associated outputs are provided. Now, the machine learning model is being trained using the unlabelled input data. It will first evaluate the raw data to identify any hidden patterns in the data before applying the appropriate algorithms. The unsupervised leaning algorithm has two subtypes: Clustering and Association.

2.2.1 Clustering

Clustering is a way of organising items into clusters so that the items that share the most similarities stay in one group and share little to none with the objects in another group. The data items are classified based on the existence or lack of commonalities discovered during the cluster analysis.

2.2.2 Association

A rule of association is used to uncover links between variables in a sizable database using unsupervised learning techniques. It establishes the group of items that co-occur in the collection. Marketing strategy is more effective because to the association rule. Those who purchase X (let us say, bread) also frequently buy Y (let us say, butter or jam). Market Basket Analysis is an illustration of an association rule in action.

Popular unsupervised learning algorithms are listed below:

A. K-means clustering; and

B. Principal component Analysis.

2.2.2.1 K-means clustering

The unlabelled dataset is divided into k different clusters using an iterative process. Each cluster comprises just one dataset and has a unique set of properties. The data objects are divided into separate clusters using an unsupervised learning algorithm, which also serves as a useful technique for automatically identifying group categories in unlabelled datasets without the need for training. As each cluster is linked to a centroid, the method is centroid-based. The main objective of this approach is to minimise the overall distances between the data points and the clusters they belong to. Because it is straightforward and linear in complexity, the K-means clustering algorithm is used for clustering WSN sensor nodes and is useful for finding the cluster heads as well.

The K-means method is demonstrated in the phases below:

1. Choose K to determine the total number of clusters;

2. Choose K points or the centroid at random in next step;

(That may not be the dataset that was provided.)

1. Assign each data point to its nearest centroid, which will produce the predetermined K clusters;

2. Locate the centroid of each cluster and calculate the variance;

3. Repeat the third step to assign each data point to its new centroid in this step;

4. Go to step 4 if a reassignment is necessary; otherwise, move to final stage;

Some pros of using K-means:

1. Easy to put into practice;

2. Big data sets are scaled;

3. Ensures convergence;

4. Can warm-up the centroids locations;

5. Adapts readily to new examples; and

6. Broadens to include clusters of various sizes and shapes, such as elliptical clusters.

Some cons of using K-means:

1. It is challenging to estimate the k-value, or the number of clusters;

2. Initial inputs like the number of clusters in a network have a significant impact on output (value of k);

3. The order in which the data is entered significantly affects the result;

4. Rescaling makes it pretty sensitive. Using normalisation or standards to rescale our data will result in a very different result. Final outcome; and

5. When clusters have a complex geometric shape, clustering activities should not be performed.

2.2.2.2 Principal component analysis (PCA)

Principal component analysis is the most effective unsupervised learning technique for reducing the dimensionality of data. It simultaneously reduces information loss while increasing interpretability. It facilitates the identification of the dataset's most crucial qualities and makes data easier to plot in 2D and 3D. PCA facilitates the discovery of a series of linear combinations of variables. The Main Components are the names given to these newly altered functions. One of the most well-known pieces of equipment for exploratory information evaluation and predictive modelling is this [20].

Typically, PCA looks for the surface with the lowest dimensionality onto which to project the high-dimensional data. PCA functions by taking into account each attribute's variance since a high attribute demonstrates a solid split between classes, which results in low dimensionality.

Since it uses a feature extraction technique, it keeps the crucial variables and discards the unimportant ones.

Some of the important additives of principal components are given below:

- The number of these components is either equal to or less than the original functions gift inside the dataset;
- The major element ought to be the linear combination of the unique capabilities;
- These components are orthogonal, i.e., the correlation between a couple of variables is zero; and
- The significance of every element decreases while going to 1 to n, it manner the 1 pc has the maximum importance, and n laptop will have the least significance.

2.2.2.3 Steps for PCA algorithm

Step 1: To obtain the dataset: First, split the input dataset into two halves, X and Y, where X represents the training set and Y represents the validation set;

Step 2: Putting information into a structure: Now create a structure to represent dataset, and use the two-dimensional matrix of independent variable X as an example. Here, each row represents a data item and each column represents a feature. The dataset's dimensions are determined by the number of columns;

Step 3: Data standardisation: Normalise the dataset in this stage. For instance, in a given column, features with higher variation are more significant than features with smaller variance. Split each piece of data in a column by the column's standard deviation if the importance of features is independent of the variance of the feature. The matrix in this case is called Z;

Step 4: Determining Z's covariance: Transpose the Z matrix in order to determine Z's covariance. Transpose it first and then multiply it by Z. The Covariance matrix of Z will be the output matrix;

Step 5: Making the Eigen Values and Eigen Vectors calculations: The resulting covariance matrix Z's eigen values and eigenvectors must now be determined. The high information axis' directions are represented by eigenvectors or the covariance matrix. Additionally, the eigen values are defined as the coefficients of these eigenvectors;

Step 6: Sorting of the Eigen Vectors: This phase involves taking all of the eigen values and sorting them from largest to lowest in a decreasing order. Additionally, in the eigen values matrix P, simultaneously sort the eigenvectors in accordance. The matrix that results in is known as P*;

Step 7: Principal Components or the new features calculation: Compute the new features here, and then multiply the P* matrix by Z to achieve this;

Step 8: Eliminate less significant or irrelevant features from the new dataset: Decide here what to keep and what to eliminate now that the new feature set has been implemented. Retain relevant or significant features in the new dataset and exclude irrelevant information.

The PCA algorithm allows the minimal variance components to be dropped because they simply have the least amount of information and reduce dimensionality. This could reduce the amount of data being communicated between sensor nodes in WSN scenarios by obtaining a small pair of uncorrelated linear combined innovative readings. By permitting the selection of only significant principle components and eliminating other lower order inconsequential components from the model, it can also turn the problem of vast data into one of tiny data.

2.3 Reinforcement learning

The enormous amount of data that models need to train on is a recurring problem in machine learning. A model may need more data the more complicated it is. After all of this, the information received might not be trustworthy. It could be inaccurate, missing, or compiled from unreliable sources. Data acquisition is solved through Reinforcement Learning, which virtually eliminates the requirement for data.

A subfield of machine learning called reinforcement learning develops a model's ability to solve issues at their best on its own. With reinforcement learning, a computer learning model must analyse the issue and find the best solution on its own. This implies that we also come up with quick and original solutions that the programmer might not have even considered. A certain class of problems, such as those in robotics, gaming, and other long-term endeavours, are solved using RL.

Key principles of reinforcement learning

- In real life, the agent is not given instructions regarding the surroundings or what needs to be done;

- It is founded on the hit-and-miss method;

- The agent performs the subsequent action and modifies its states in response to feedback from the preceding action;

- The agent might receive a reward afterwards; and

- The agent must investigate the stochastic environment in order to maximise positive rewards.

Reinforcement learning's advantages

- Reinforcement learning can be used to resolve extremely complicated issues that cannot be resolved using traditional methods;
- To attain long-term effects, which are very challenging to achieve, this method is favoured;
- The way that humans learn is remarkably similar to this learning model;
- The model is capable of fixing mistakes that happened during training;
- The likelihood of experiencing the same error after a model has rectified one is quite low; and
- To tackle a certain issue, it might produce the ideal model.

The drawbacks of reinforcement learning

- In many ways, the framework of reinforcement learning is flawed, but it is precisely this flaw that makes it valuable.
- A state overload brought on by excessive reinforcement learning may have a negative impact on the outcomes.
- Using reinforcement learning to solve straightforward issues is not recommended.
- Both a lot of data and a lot of compute are required for reinforcement learning. It craves information. Because one can play video games repeatedly, gathering a lot of data looks doable; this is why it works so well in them.

Algorithms for reinforcement learning can be used by robots to teach themselves to walk. The mainly used algorithm in reinforcement learning is: **Q learning.**

2.3.1 Q learning

Given the agent's present state, Q-learning [21] is a model-free, off-policy reinforcement learning technique that will determine the appropriate course of action. The agent will choose what to do next base on its location in the surroundings. The model's goal is to determine the optimum course of action given the situation as it is. In order to accomplish this, it might devise its own set of rules or might deviate from the prescribed course of action. This indicates that there is no real need for a policy, which is why it is referred to as off-policy. Model-free refers to the use of predictions about the environment's anticipated response by the agent in order to make decisions. It instead relies on trial and error learning rather than the reward system.

A recommendation system for advertisements is an illustration of Q-learning. The advertisements you see in a typical ad suggestion system are determined by your past purchases or websites you may have visited. In the event that you have already purchased a TV, various brand TVs will be suggested. A distributed architecture such as a wireless sensor network, where each node conducts activities that are anticipated to optimise its long-term benefits can readily implement this algorithm.

3. Conclusion

Due to the numerous ways that wireless sensor networks differ from regular networks, there is a need for protocols and tools that address particular problems and constraints. This chapter provided an empirical study of the wireless sensor network infrastructure, including its architecture, applications, and limitations. These networks consequently need novel approaches to routing, security, scheduling, localization, node clustering, data aggregation, fault detection, and data integrity that are both energy conscious and real-time. After that, the Chapter provided the taxonomy of ML algorithms that were applied to WSNs and performed a quantitative analysis of the algorithms that helped WSNs to overcome those limitations. Furthermore, varieties of methods are offered to improve a wireless sensor network's capacity to adjust to the changing behaviour of its environment. We also highlighted each ML algorithm's benefits and drawbacks.

However there is still a lot of work being done to address numerous outstanding issues because applications of machine learning methods to wireless sensor networks are still a relatively new area of study.

Author details

Rakesh Chandra Gangwar* and Roohi Singh
Department of Computer Science and Engineering, Sardar Beant Singh State University, Gurdaspur, Punjab, India

*Address all correspondence to: rakeshgangwar@sbssugsp.ac.in

References

[1] Akyildiz IF, Weilian S, Sankarasubramaniam Y, Cayirci E. A survey on sensor networks. IEEE Communications Magazine. 2002;**40**(8): 102-114

[2] Romer K, Mattern F. The design space of wireless sensor networks. IEEE Wireless Communications. 2004;**11**(6): 54-61

[3] Kalantary S, Taghipour S. A survey on architectures, protocols, applications and management in wireless sensor networks. Journal of Advanced Computer Science & Technoloy. 2014;**16**:1-11

[4] Đurišić MP, Tafa Z, Dimić G, Milutinović V. A survey of military applications of wireless sensor networks. In: Proceedings of the 2012 Mediterranean Conference on Embedded Computing (MECO); Bar, Montenegro: IEEE (Piscataway); 2012. pp. 19-21

[5] Bokareva T, Hu W, Kanhere S, Ristic B, Gordon N, Bessell T, et al. Wireless sensor networks for battlefield surveillance. In: Proceedings of the Land Warfare Conference. Brisbane, Australia: MDPI (Switzerland); 2006. pp. 1-8

[6] Hii P, Chung W. A comprehensive ubiquitous healthcare solution on an android mobile device sensors. Sensors (Basel). 2011;**11**(7):6799-6815

[7] Nasir A, Soong BH, Ramachandran S. Framework of WSN based human centric cyber physical in-pipe water monitoring system. In: Proceedings of the 2010 11th International Conference on Control Automation Robotics & Vision; Singapore: IEEE (Piscataway); 2010. pp. 7-10

[8] Mansour S, Nasser N, Karim L, Ali A. Wireless sensor network-based air quality monitoring system. In: Proceedings of the 2014 International Conference on Computing, Networking and Communications (ICNC), 3-6 February; Honolulu, HI, USA: IEEE (Piscataway); 2014. pp. 545-550

[9] Khedo KK, Bissessur Y, Goolaub DS. An inland wireless sensor network system for monitoring seismic activity. Future Generation Computer Systems. 2020;**105**:520-532

[10] Kassim M, Harun AN. Applications of WSN in agricultural environment monitoring systems. In: Proceedings of the 2016 International Conference on Information and Communication Technology Convergence (ICTC), 19-21 October; Jeju, Korea: IEEE (Piscataway); 2016

[11] Akyildiz IF, Su W, Sankarasubramaniam Y, Cayirci E. Wireless sensor networks: A survey. Computer Networks. 2002;**38**(4): 393-422

[12] Praveen KD, Amgoth T, Annavarapu CSR. Machine learning algorithms for wireless sensor networks: A survey. Information Fusion. 2019;**49**: 1-25

[13] Javaid A, Javaid N, Wadud Z, Saba T, Sheta OE, Saleem MQ, et al. Machine learning algorithms and fault detection for improved belief function based decision fusion in wireless sensor networks. Sensors. 2019;**19**:1334

[14] Di M, Joo EM. A survey of machine learning in wireless sensor networks from networking and application perspectives. In: 6th International Conference on Information, Communications Signal Processing. New York: Springer; 2007. pp. 1-5

[15] Sarker IH, Colman A, Han J, Khan AI, Abushark YB, Salah K. BehavDT: A behavioral decision tree learning to build user-centric context-aware predictive model. Mobile Networks and Applications. 2020;**25**: 1151-1161

[16] Lee SM, Kim DS, Park JS. A hybrid approach for real-time network intrusion detection systems. IEEE Transactions on Vehicular Technology. 2011;**60**:457-472

[17] Pisner DA, Schnyer DM. Machine Learning: Support Vector Machine. Amsterdam: Elsevier; 2020. pp. 101-121

[18] Dey A. Machine learning algorithms: A review. International Journal of Computer Science and Information Technologies (IJCSIT). 2016;**7**(3): 1174-1179

[19] A. Forster, "Machine learning techniques applied to wireless ad-hoc networks: Guide and survey." In 3rd International Conference on Intelligent Sensors, Sensor Networks and Information. IEEE, pp. 365-370, 2007.

[20] Feldman D, Schmidt M, Sohler C, Feldman D, Schmidt M, Sohler C. Turning Big Data into Tiny Data: Constant Size Coresets for k-Means, PCA and Projective Clustering. New Orleans, USA: SODA-2013; 2013. pp. 1434-1453

[21] Watkins C, Dayan P. Q-learning. Machine Learning. 1992;**8**(3-4):279-292

Chapter 4

Efficient Machine Learning Classifier for Fault Detection in Wireless Sensor Networks

Poornima G. Miathali

Abstract

The deployment of wireless sensor networks in unpredictable and dangerous conditions makes them prone to software, hardware, and communication errors. Sensors are physical devices that are deployed in inaccessible environment which makes them malicious. The Fault occurs in the sensed data and its detection should be precise and rapid to limit the loss. The status of sensed data should be explicitly determined to guarantee the normal function of the Sensor Networks. For the purpose of fault detection machine learning classifiers are employed because they are effective and used to classify sensed data into faulty and non-faulty data. The faults due to Dos, Probe, R2L, and U2R are considered for implementation. KDD CUP 99 dataset is chosen for training and test purpose, and the dataset contains 41 features which are categorized as content, basic, TCP features. The required feature for each fault category is selected through recursive feature elimination technique. The performance of the classifier is measured and evaluated in terms of Accuracy, precision, recall, and F-measures. From experimental results, it is observed that Random Forest classifier is best suited for Wireless Sensor Networks fault detection. The simulation result shows that Multi-layer perceptron outperforms the other classifier with 92% of accuracy.

Keywords: attacks, classifiers, sensor networks, machine learning, random forest, support vector machine, multilayer perceptron, stochastic gradient descent, faults

1. Introduction

Wireless sensor networks are widely used for a variety of purposes, including systems that must function safely. More mission-critical subsystems like cars, drones, and others are joining the area of WSNs, although historically, geographically close systems would link wirelessly over time. As a result, it has become imperative to create WSNs that are fault tolerant. Data security plays a crucial part in successful communication. In earlier days for the security purpose Encryption, Firewall, Virtual Private networks (VPN) were used to provide data security. But these methods are not enough to secure the data. Therefore, machine learning approach gives an effective way to deal with the problem. Many researches have performed studies and arrived at various conclusions on data safety. With hardware implementation, these methods seem to be complex. So, machine learning gives the best solution for the problem. This is the easiest way and does not consume large amount of time compared to other methods and at the same time it is a cost-effective method.

From Education to Entertainment industry data is the backbone. Therefore data security and safety is significant. Hackers may duplicate the data packets or even IP address itself therefore it is difficult to identify the malicious data in the network. Machine learning techniques give the efficient solution.

A Hardware model is implemented [1], using sensors. This method seems to be complicated. Attack detection has achieved through block chain technology [2], But this method suffers from computational delay, block chain overheads, cost of implementations. Other machine learning classifiers are used to find the attacks. The major drawback from the research is more computational time and more false positive values [3, 4]. In the present study false positive values are comparatively less and it is shown in the confusion matrix and it is discussed in the result section.

The Data set considered for the present study is KDD Cup 99 dataset which contains large number of data sets and is publically available. The major attacks that are considered in this attack are DoS (Denial of Service) attack in which an unauthorized user getting access to the network. Probe attack. R2L (Remote to user) attack in which an unauthorized user can send data packets to the system where he or she cannot have the access as a local user. U2R (User to root) attack in which the unauthorized can get into the root.

To analyze the data as attacked or normal data four classifiers are considered, RF (Random Forest), Support Vector machine (SVM), Multilayer Perceptron (MLP), Stochastic Gradient Descent (SGD) Classifiers are used. Raw data cannot be used to test and train the machine learning model. So, Data preprocessing steps such as Feature Selection, Encoding. The Preprocessed data are applied to the different classifiers. Efficiency parameters such as accuracy, precision, recall, F-measure, selectivity, specificity, G-mean are found out. By comparing all these parameters the final result can be achieved. Different percentage of attacks can be introduced in the data set. So that an efficient classifier can be found out for different percentage of attacks. The Efficiency parameter can be obtained from Confusion matrix. Confusion matrix contains True positive (TP), True Negative (TN), False Positive (FP), False Negative (FN) value.

In the present study, a brief description on available data set in the internet is presented. Further, pre-processing of data is discussed in detail. When data sets are applied to 4 different types of classifiers, efficient classifier is derived with respect to confusion matrix parameter.

The paper is organized as follows, Section 2 the motivation for the present study is discussed, Section 3 reviews the related works carried out in the field of intrusion detection system and various data faults that occur and the type of classifiers used is presented. Section 4 introduces the proposed Method, In Section 5 discusses the performance measures and analyses. The Paper finally concludes with Section 6 with future research directions.

2. Motivation

Wireless sensor networks are widely used for a variety of purposes, including systems that must function safely. More mission-critical subsystems like cars, drones, and others are joining the area of WSNs, although historically, geographically close systems would link wirelessly over time. As a result, it has become imperative to create WSNs that are fault tolerant. The nature of the defects that are likely, as described in the introduction section, makes it appropriate to cutting-edge technology, such as machine learning, to find such faults.

3. Literature survey

The focus of the research work presented in this paper is on the detection of faults due to attacks and the methods used to detect and classify the data.

Zainib Noshad, Nadeem Javaid, Tanzila Saba [1] The use of wireless sensor networks (WSNs) in a variety of environments makes them susceptible to errors. Unstable and dangerous conditions. This makes WSN vulnerable to errors in software, faults in hardware and communication. Fault detection in WSNs has become a challenging task because of the sensor's constrained resources and varied deployment environments. The classification of gain, offset, spike, data loss, out of bounds, and stuck-at faults at the sensor level is done using Support Vector Machine (SVM), Convolutional Neural Network (CNN), Stochastic Gradient Descent (SGD), Multilayer Perceptron (MLP), Random Forest (RF), and Probabilistic Neural Network (PNN) classifiers. Two of the six faults—the spike and data loss faults—are brought on by the datasets. The Detection Accuracy (DA), True Positive Rate (TPR), Matthews Correlation Coefficients (MCC), and F1-score are used to compare the results. Simulations demonstrate that the RF method achieves a higher rate of defect detection than the other classifiers.

Salah Zidi, Tarek Moulahi, and Bechir Alaya [2] one of the easiest ways to find failure in WSNs appears to be to use machine learning. SVM is employed in our context to define a decision function, which is based on statistical learning theory. This technique has a lot of potential for multidimensional data learning in addition to having demonstrated performance in a number of fields. This method, which makes use of kernel functions, has a significant capacity for adaptability for nonlinear classification scenarios, such as our case of fault detection. This has the potential to be very helpful in fault prevention. The goals of this research are to use a dynamic classification approach to track sensor activity through its data in order to predict errors as quickly as feasible in the same context of prevention.

Terry Windeatt [3] Multilayer Perceptron (MLP) classifier settings can be difficult to adjust, as is widely known. In this study, a metric that can forecast how many classifier training iterations will take to get the best results from an ensemble of MLP classifiers is described. The measure, which is based on a spectral representation of a Boolean function, is computed between pairs of patterns on the training data. With this representation, accuracy and diversity can be combined into a single statistic that describes the mapping from classifier decisions to the target label.

Luofan Dong, Huaqiang Du, Fangjie Mao [4] Convolutional neural networks (CNNs) recently demonstrated outstanding performance in a variety of applications, including computer vision and remote sensing semantic segmentation. Much interest is focused on the capacity to learn CNN's high-representational properties. On the other hand, the random forest (RF) technique is frequently used for variable selection, classification, and regression. This article tested a technique based on the fusion of an RF classifier and the CNN for a very high-resolution remote sensing (VHRRS) based forests mapping. This method was based on the previous fusion models that fused CNN with the other models, such as conditional random fields (CRFs), support vector machine (SVM), and RF. Huwaida T. Elshoush, Esraa A. Dinar [5] Spam prevalence is rising daily as electronic emails are used more frequently. As a result, spam emails have grown to be a serious issue that reduces the use of electronic emails for communication. Several machines learning approaches, including Naive Bayes (NB), K-Nearest Neighbor (KNN), Support Vector Machine (SVM), Artificial Neural Network (ANN), and Decision Tree, provide email spam filtering solutions (DT). This study examines various machine learning methods, namely Adaboost and Stochastic Gradient Descent, to filter spam emails (SGD).

Mrutyunjaya Panda, Ajith Abraham [6] Security of network traffic is growing to be a significant issue for computer networks as the internet expands. The frequency of attacks on the network is rising over time. Such network attacks are nothing more than intrusions. The network and data have been protected against threats by using intrusion detection systems to identify intrusions. Large amounts of network data are monitored, analyzed, and classified into abnormal and regular data using data mining algorithms. Poornima G, K Suresh Babu, K B Raja, K R Venugopal, and L M Patnaik [7] proposed to find the probability of correctly identifying a faulty node for three different types of faults based on normal bias. The nodes fault status is declared based on its confidence score that depends on the threshold valve. Uma R. Salunkhe, Suresh N. Mali [8] used an intrusion detection system (IDS) to detect hostile activity has been an efficient technique to increase security. An intrusion detection system is anomaly detection. Due to its inability to accurately detect all sorts of attacks, current anomaly detection is frequently characterized by high false alarm rates and only modest accuracy and detection rates. Using the KDD-99 Cup and NSL-KDD datasets, a test is run to assess how well the various machine learning methods perform.

Miao X, Liu Y, Zhao H, Li C [9] system which detects the attacks in the wireless sensor network the KDD cup 99 data set is used in the present paper and the to classify the attacks in the WSN's the KNN classifier is used, But the detection rate achieved with this classifier is very poor and the highest detection rate is 75% and that is for k = 5. Gharghan S.K, Nordin R, Ismail M, Ali J. A [10] a hardware model for intrusion detection system is suggested this model has failed to give the accurate result, due to some hardware vulnerabilities and it is complex to design and human intervention is required.

In [11] the authors have discussed Intrusion detection system and used Decision tree, SVM, MLP algorithm. The result shows that MLP outperforms the other classifier with accuracy of 91%. In [12] the authors elaborate on layer wise DoS attack and its defense mechanisms and classification. In [13] the authors detect faults in WSN using hidden Markov model, KDD cup 99 data set is used, the accuracy they have achieved for test data is 77.11%. In paper [14] fault detection using deep learning algorithms is done. KDD cup 99 data set is used and MLP, SVM algorithms are used and the accuracy is 91%. In [15] the fault detection in WSN using Internet of things based on improved BP Neural network Leven berg- Marquard algorithm is applied with a accuracy result of 91%.

From the papers surveyed, for selecting feature subset Recursive feature elimination method is implemented. All the independent variables in supervised learning is known as features of the data. Elimination in this context means eliminating the features. Doing a process repetitively to eliminate the features of the data is known as Recursive feature elimination. KDD is a type of data set and an online repository that contains data from all different sorts of intrusion attempts. It mainly includes DOS, R2L, U2R, and PROBE. RF, SVM, SGD, MLP classifiers will be assessed on the KDD dataset in this research.

4. Methodology

The methodology used in this work is shown in the **Figure 1** and it involves preprocessing the KDD dataset initially, using the prepared dataset in a fair environment with equal access to resources, and then comparing classifier performance across all analyzed attacks (DOS, R2L, U2R, and PROBE) and their faults. Machine learning model needs large number of data set to avoid the problem of over fitting. The Proposed optimal feature subset selection algorithm includes feature normalization, feature scoring, feature subset selection and feature subset elimination.

Data Preprocessing is the most time-consuming task but plays significant role in machine learning model. Raw data cannot be used for training or testing the machine learning model. Hence data preprocessing is required in machine learning. Encoding is the process of converting the categorical data into numerical value. Categorical values are the string values that are stored as components of the input features. Features/attributes that have strings or categories as their values are termed as categorical attributes. These Categorical values can be represented in two forms, namely Nominal and Ordinal. When there is no ordering between the attributes those are referred to as Nominal attributes. When there is an ordering between the attributes those are referred to as Ordinal attributes. KDD cup 99 data set contains 125,973 train data and 22,544 test data. So, it helps to build and test an efficient machine learning model. It also contains different type of attacks called Neptune, pod, smurf, etc. which can be further categorized as DoS, Probe, U2R and R2L attacks [16] as shown in **Table 1**.

In NSL-KDD cup 99 data set with 41 input features are present. In that protocol feature contains tcp, udp, icmp etc., Service feature contains http, ftp, telnet etc., Flag feature contains SF, REL, ROIT etc. These three columns contain symbolic and continuous data which cannot be used. Because the classifier that are considered, accepts only numerical values. Hence One Hot Encoding technique is used. One hot encoding columns are exactly equal to the number of the values a particular feature is having. While encoding, there should be only one value present only once in the encoded values.

Feature Scaling is the process of transforming the data value into 0 to 1. For example if we consider two weights whose values are 80 and 40 respectively, by

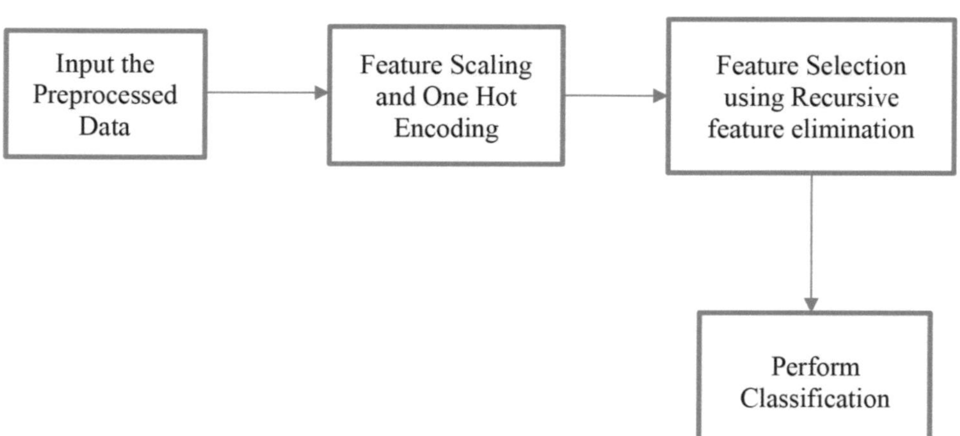

Figure 1.
Block diagram of the proposed method.

Attack types	DoS attack	Probe attack	U2R attack	R2L attack
Known attack	Neptune, back, land, pod, smurf teardrop	Ipsweep, nmap, portsweep, satan	ftp_write, guess_ password, imap, multihop, phf, spy, wazerclient, wazermaster	Buffer_ov erflow, landmod ule, perl, rootkit
New attack	Mailbomb, apache2, processtable, worm	mscan, saint	Sendmail, snmpattack, snmpguess, httptunnel	Ps, sqlattack, xterm

Table 1.
Types of attacks and faults in the data.

feature scaling these can be represented as 1 and 0 where 0 is the lowest possible score/weight and 1 is the highest possible score/weight.

Feature Selection: Machine learning model needs to be trained by huge number of data set for the accurate result. But some data does not contain useful information, without considering that feature also classification can be done. This process is known as Feature selection. Feature Selection basically a process where in only few features that contains the useful information can be selected and machine learning model will get rid of the noise data. Recursive feature elimination method is used for feature selection. All the independent variables in supervised learning is known as features of the data. Elimination in this context means eliminating the features. Doing a process repetitively to eliminate the features of the data is known as Recursive feature elimination (RFE). RFE works by searching for a subset of features by starting with all features in the training dataset and successfully removing features until the desired number remains. KDD cup 99 data set has 41 input features. Among 41 features, 23 features are selected for the model by using the recursive feature elimination technique (**Table 2**).

Classifiers that are used to classify the malicious and the normal data are:

1. Random Forest Algorithm [17]: Decision Trees are very sensitive by nature, necessitating the use of the Random Forest algorithm. The decision tree's entire structure may change if the training data set has a slight difference. Because of this, it is extremely sensitive, and the outcome is highly variable. Decision trees are the binary trees that recursively splits the data set until we are left with pure

Selected features
src_ bytes; Dst_bytes; Wrong_fragment; Num_compromised; Count; Srv_count;
Same_srv_rate; Diff_srv_rate; Dst_host_same_srv_rate; Dst_host_error_rate;
Dst_host_srv_error_rate; Protocol_type; Dst_host_diff_srv_rate; Dst_host_same_src_port_rate
Dst_host_srv_diff_host_rate; Service_eco_i; Hot; Logged_in
Is_guest_login; Dst_host_count; Dst_host_srv_count

Table 2.
Selected features.

leaf nodes. Bootstrapping is the process of building a new data set from an existing one. We must use the bootstrapped data sets to train the decision trees. This is how the data is aggregated using Random Forest Classifier.

2. Support Vector machine: SVM classifier comes under the supervised learning. When the data is of 2- dimensional space then a line which separates the two classes needs to be created. But the data set is in 3 dimensional then a hyper plane that will separate the 3-dimensional data sets needs to be created. a line can be used to demarcate two-dimensional linearly separable data. Y = Ax+B is the definition of the line's function. The equation becomes x2 = ax1 + b if we replace x in this case with x1 and y with x2. The new form of this equation is ax1-x2 + b = 0. We will obtain wx + b = 0 if we define x = (x1, x2) and w = (a,-1). Thus, we shall obtain the line's equation. The same line is used to divide the two data classes in logistic regression as well, however the problem with logistic regression is that it does not care if the cases are actually close to the line or not.

3. Multilayer Perceptron: A fully connected class of feed forward artificial neural network is called a multilayer perceptron (MLP) (ANN). The term "MLP" is used ambiguously; sometimes it is used broadly to refer to any feed forward ANN, and other times it is used specifically to describe networks made up of several layers of perceptron's (with threshold activation). Especially when they comprise a single hidden layer, multilayer perceptron's are commonly referred to as "vanilla" neural networks in common parlance. In MLP input layer, output layer and number of hidden layers are used depending on weights and activation function the classification is done in output layer. Each node, with the exception of the input nodes, is a neuron that employs a nonlinear activation function. Back propagation is a supervised learning method that is used by MLP during training. MLP differs from a linear perceptron [18] due to its numerous layers and non-linear activation. It can identify non- linearly separable data.

4. Stochastic gradient descent: It is an iterative process to optimize the objective function. Gradient simply refers to a surface's slope or tilt. Gradient descent is an iterative process that descends a function's slope in steps from a random point until it reaches the function's lowest point.

Performance Evaluation Measures: In this section, we provide a detailed evaluation of the machine learning techniques with various performance measures to detect network faults caused due to intrusions.

4.1 Confusion matrix

Confusion matrices are a widely used measurement when attempting to solve classification issues. Both binary classification and multi class classification issues can be solved with it. In Confusion matrix there are values which are called True Positive, True Negative, False Positive and False Negative. True Positive Constitutes the data features that are correctly identified by the Algorithm. True Negatives are also the values that are correctly identified by the algorithm. False Positive and the False Negative are the data features that are wrongly identified by the Algorithm. The Confusion matrix in machine learning is used to find the Precision and Accuracy of the Classifier which we can obtain those from True and False Values. After the classifiers are trained the performance of all 4 classifiers are measured in terms of these metrics using test data set. Based on the Confusion Matrix, Accuracy, Precision, Recall, F-measure, Specificity, Selectivity, G-mean are calculated as mentioned below,

1. **Accuracy**: Accuracy of a classifier can be calculated as ratio total true values with all the values present in the confusion matrix.

$$\text{Accuracy} = (\text{TP} + \text{TN})/(\text{TP} + \text{TN} + \text{TP} + \text{TN}) \tag{1}$$

2. **Precision**: Precision is determined by dividing the total number of optimistic predictions by the actual number of optimistic predictions.

$$\text{Precision} = (\text{TP})/(\text{TP} + \text{FP}) \tag{2}$$

3. **Recall**: It is obtained by dividing the sum of all valid samples by the total number of valid positive predictions.

$$\text{Recall} = (\text{TP})/(\text{TP} + \text{FN}) \tag{3}$$

4. **F-measure:** The F1 score is defined as the harmonic mean of precision and recall.

$$\text{F} - \text{measure} = 2^* \ (\text{Precision}^* \ \text{Recall})/(\text{Precision} + \text{Recall}) \tag{4}$$

5. **G-mean**: Geometric mean is the square root of true positive rate and true negative rate.

$$\text{Gmean} = \boldsymbol{TPR}^* \ \boldsymbol{TNR} \tag{5}$$

6. **Selectivity**: Sensitivity is the ratio between the total number of positive samples to the number of samples tested as positive in the test

$$\text{Sensitivity} = (\text{TP})/(\text{TP} + \text{FN}) \tag{6}$$

7. **Specificity**: Specificity is the ratio between total numbers of negative samples to the number of samples tested as negative in the test.

$$\text{Specificity} = (\text{TN})/(\text{FP} + \text{TN}). \tag{7}$$

5. Result and discussion

In this section, the experimental results of our machine learning techniques with four class classification methodology using NSL-KDD intrusion detection dataset are provided in order to detect network intrusions and then comparison with the existing approaches is done to evaluate the efficacy of our network intrusion detection model. Confusion matrix is drawn for each type of attack and their faults of all four classifiers. So we obtain 16 set of confusion matrix a n d which are presented in this study. The performance of the classifiers is measured in terms of Confusion matrix,

Accuracy, Precision, Recall, F-measure, Specificity, Selectivity, G-mean. After the classifiers are trained the performance of all 4 classifiers are measured in terms of these metrics using test data set.

All the experiments are conducted using NSL-KDD dataset that has 125,973 training instances, 22,544 instances for testing with 41 attributes and 4 attack types for four classifiers to build an efficient network fault detection system. We have evaluated all algorithms with various evaluation measures, as discussed in the above section.

Confusion matrix for Random Forest:

Confusion matrix for major types of faults is shown in the **Figures 2–5**. For U2R attack the True negative is zero since the fault data is very low compared to the normal data, In R2L also number of fault data is very low, therefore the true negative value is low. In DoS and Probe attack also number of False positive data is more therefore the accuracy will be less. In the similar way the Confusion matrix for other classifiers are also constructed.

Tables 3–6 show the performance of the all 4 classifiers and from the result obtained we see the MLP classifier performs better than the other Classifier. False positive rate is less for MLP Classifier, True Positive and True Negative values are more. Therefore, MLP classifier is efficient classifier for fault detection Wireless Sensor Networks. The comparative plot of Accuracy for all 4 types of classifier algorithm is shown in the **Figure 6**, and it's evident that MLP on an average has an accuracy of 89.725%.

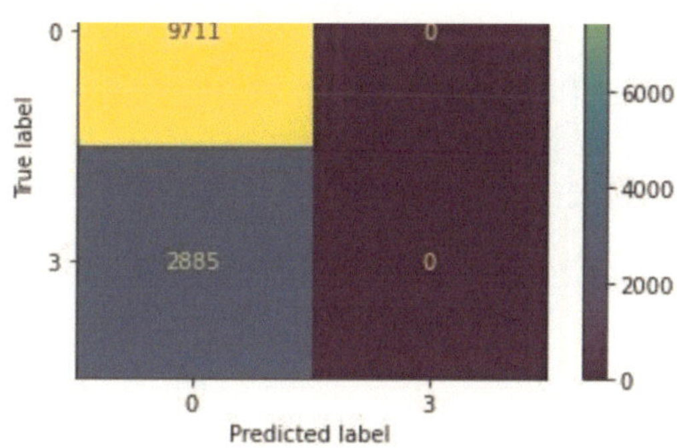

Figure 2.
U2R attack for RF.

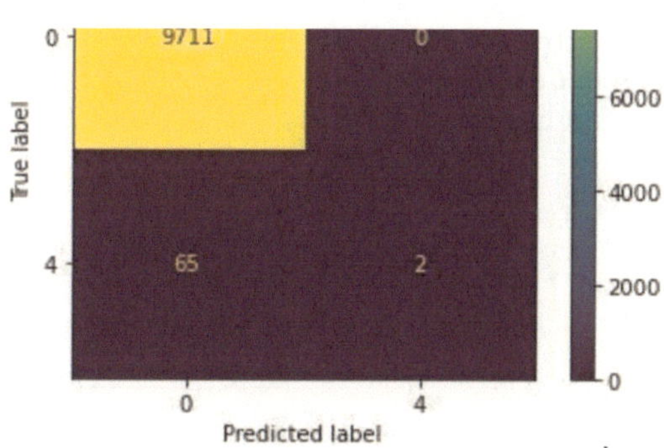

Figure 3.
R2L attack for RF.

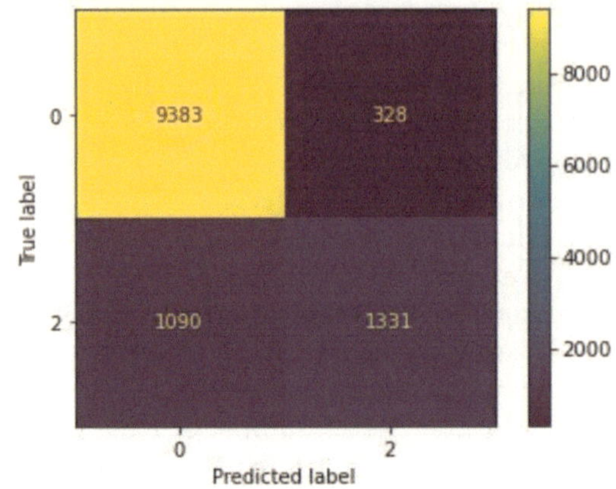

Figure 4.
Probe attack for RF.

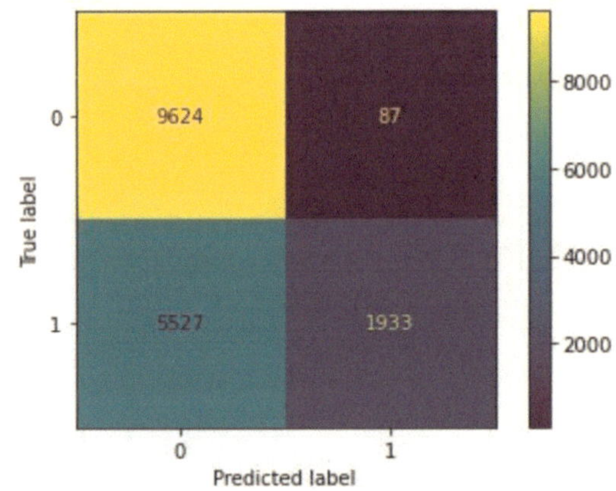

Figure 5.
DoS attack for RF.

The performance of the classifiers is also studied by introducing different Fault Probability Rates (FPR) and the results of the same are shown in **Tables 7–10**. The major goal of the fault percentage variation is that how accurately a classifier classifies the attack or normal data irrespective of the percentage of the fault present in particular test data. In the present study a classifier classifies the data with good amount of accuracy even if the percentage of fault is high.

Attack Types	Accuracy	Precision	Recall	F-measure	Specificity	Selectivity	G-mean
DoS	0.880	0.828	0.995	0.904	0.003	1.00	0.05
Probe	0.867	0.885	0.957	0.920	0.508	0.975	0.70
R2L	0.770	0..770	1.000	0.870	0.000	1.000	0.00
U2R	0.993	0.993	1.000	0.996	0.000	1.000	0.00

Table 3.
Accuracy for random Forest classifier.

Attack Types	Accuracy	Precision	Recall	F-measure	Specificity	Selectivity	G-mean
DoS	0.868	0.820	0.983	0.894	0.720	0.983	0.84
Probe	0.876	0.900	0.950	0.925	0.579	0.950	0.74
R2L	0.993	0.993	1.000	0.996	0.000	1.000	1.00
U2R	0.771	0.771	0.999	0.870	0.000	0.998	0.02

Table 4.
Accuracy for support vector machine classifier.

Attack Types	Accuracy	Precision	Recall	F-measure	Specificity	Selectivity	G-mean
DoS	0.886	0.842	0.982	0.907	0.782	0.981	0.87
Probe	0.920	0.935	0.935	0.935	0.591	0.969	0.75
R2L	0.993	0.994	0.998	0.996	0.000	0.999	0.00
U2R	0.790	0.790	0.998	0.880	0.265	0.999	0.51

Table 5.
Accuracy for MLP classifier.

Attack Types	Accuracy	Precision	Recall	F-measure	Specificity	Selectivity	G-mean
DoS	0.885	0.895	0.903	0.899	0.851	0.900	0.87
Probe	0.810	0.812	0.992	0.893	0.663	0.992	0.25
R2L	0.993	0.993	1.000	0.996	0.000	1.000	0.00
U2R	0.771	0.771	0.999	0.870	0.001	0.001	0.04

Table 6.
Accuracy for SGD classifier.

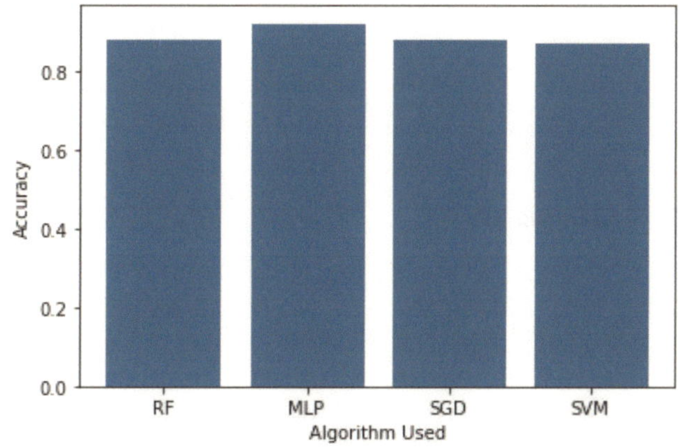

Figure 6.
Comparison of accuracy for all four classifiers.

Efficiency Parameter	FPR = 0.5	FPR = 0.4	FPR = 0.3	FPR = 0.2	FPR = 0.1	FPR = 0.05
Accuracy	0.880	0.750	0.834	0.932	0.873	0.978
Precision	0.828	0.823	0.830	0.964	0.994	0.991
Recall	0.995	0.969	0.996	0.962	0.876	0.998
F-measure	0.904	0.892	0.875	0.961	0.931	0.988
Specificity	0.993	0.968	0.962	0.997	0.930	0.993
Selectivity	0.011	0.523	0.684	0.652	0.781	0.669
G-mean	0.167	0.711	0.811	0.800	0.852	0.815

Table 7.
Performance of RF for varying fault rate.

Efficiency Parameter	FPR = 0.5	FPR = 0.4	FPR = 0.3	FPR = 0.2	FPR = 0.1	FPR = 0.05
Accuracy	0.882	0.926	0.940	0.936	0.926	0.914
Precision	0.906	0.976	0.992	0.979	0.997	0.999
Recall	0.953	0.946	0.942	0.935	0.926	0.937
F-measure	0.929	0.961	0.966	0.965	0.960	0.974
Specificity	0.999	0.999	0.979	0.949	0.824	0.819
Selectivity	0.999	0.993	0.973	0.953	0.884	0.821
G-mean	0.999	0.993	0.973	0.933	0.829	0.799

Table 8.
Performance of SVM for varying fault rate.

Efficiency Parameter	FPR = 0.5	FPR = 0.4	FPR = 0.3	FPR = 0.2	FPR = 0.1	FPR = 0.05
Accuracy	0.894	0.922	0.944	0.961	0.944	0.956
Precision	0.905	0.947	0.963	0.981	0.998	0.982
Recall	0.964	0.966	0.976	0.978	0.945	0.972
F-measure	0.936	0.956	0.969	0.979	0.970	0.977
Specificity	0.782	0.750	0.794	0.779	0.787	0.790
Selectivity	0.981	0.975	0.980	0.986	0.979	0.985
G-mean	0.876	0.855	0.882	0.877	0.878	0.882

Table 9.
Performance of MLP for varying fault rate.

6. Conclusions

The proposed system uses different Machine learning classifiers to recognize and categorize faults in Wireless Sensor networks. The dataset has four major classes, they are DoS, Probe, R2L, U2R which are further categorized. In this paper for the purpose of fault detection Random Forest (RF), Support Vector Machine (SVM), Stochastic

Efficiency Parameter	FPR = 0.5	FPR = 0.4	FPR = 0.3	FPR = 0.2	FPR = 0.1	FPR = 0.05
Accuracy	0.880	0.758	0.647	0.734	0.944	0.939
Precision	0.887	0.932	0.924	0.986	0.994	0.998
Recall	0.900	0.717	0.609	0.720	0.984	0.940
F-measure	.893	0.810	0.734	0.833	0.999	0.968
Specificity	0.851	0.864	0.634	0.661	0.757	0.358
Selectivity	0.900	0.717	0.963	0.950	0.945	0.940
G-mean	0.875	0.787	0.782	0.793	0.846	0.886

Table 10.
Performance of SGD for varying fault rate.

Gradient Descent (SGD), Multi-layer Perceptron (MLP) a classifiers are used to classify sensed data into faulty and non-faulty data Fault detection is a challenging task since wireless networks are placed in confined spaces. Machine learning classifiers are employed in this project because they are effective. The ML algorithms are trained using preprocessed data sets. One Hot Encoding is the method that is used to pre-process the data. Since in the data set few columns does not contain Numeric values. Recursive feature elimination is used to select the features that are applicable and which helps to find the specific attack. The system is put to the test on data set that were not seen during the training phase, some new attacks are introduced in the test data and the result show that the system is effective in identifying faults in the WSN. Since fault detection in the WSN can be challenging, due to harsh environment where the WSN are deployed makes them vulnerable to faults. Therefore machine learning is essential for fault detection since it is less time consuming, faster and also gives the good accuracy.

Author details

Poornima G. Miathali
Department of Electronics and Communication Engineering, BMS College of Engineering, India

*Address all correspondence to: gpoornima.ece@bmsce.ac.in

References

[1] Noshad Z, Javaid N, Saba T, Wadud Z, Saleem MQ, Alzahrani ME, et al. Fault detection in wireless sensor networks through the random Forest classifier. Sensors. 2019;**19**:1568. DOI: 10.3390/s19071568

[2] Zidi S, Moulahi T, Alaya B. Fault detection in wireless sensor networks through SVM classifier. IEEE Sensors Journal. 2018;**18**(1):340-347. DOI: 10.1109/JSEN.2017.2771226

[3] Windeatt T. Accuracy/diversity and ensemble MLP classifier design. IEEE Transactions on Neural Networks and Learning Systems. 2006;**17**(5):1194-1211. DOI: 10.1109/TNN.2006.875979

[4] Elshoush HT, Dinar EA, Using Adaboost and Stochastic gradient descent (SGD) Algorithms with R and Orange Software for Filtering E-mail Spam. In: 2019 11th Computer Science and Electronic Engineering (CEEC). Colchester, UK: IEEE; 2019. pp. 41-46. DOI: 10.1109/CEEC47804.2019.8974319

[5] Dong L et al. Very high-resolution remote sensing imagery classification using a fusion of random Forest and deep learning technique—Subtropical area for example. IEEE Journal of Selected Topics in Applied Earth Observations and Remote Sensing. 2020; **13**:113-128. DOI: 10.1109/ JSTARS.2019.2953234

[6] Jazzar M, Yousef RF, Eleyan D. Evaluation of machine learning techniques for email spam classification. International Journal of Education and Management Engineering (IJEME). 2021;**11**(4):35-42. DOI: 10.5815/ ijeme.2021.04.04

[7] Panda M, Abraham A. Hybrid evolutionary algorithms for classification data mining. Neural Computing and Applications. 2015;**26**:507-523. DOI: 10.1007/s00521-014-1673-2

[8] Poornima G, Suresh Babu K, Raja KB, Venugopal KR, Patnaik LM. Fault diagnosis approach for WSN using Normal bias technique. ACEEE International Journal on Communication. 2013;**4**(2):29-36

[9] Zidi S, Moulahi T, Alaya B. Fault detection in wireless sensor networks through SVM classifier. IEEE Sensors Journal. 2017;**18**:340-347

[10] Muhammed T, Shaikh RA. An analysis of fault detection strategies in wireless sensor networks. Journal of Network and Computer Applications. 2017;**78**:267-287

[11] Miao X, Liu Y, Zhao H, Li C. Distributed online one-class support vector machine for anomaly detection over networks. IEEE Transactions on Cybernetics. 2018;**99**:1-14

[12] Gharghan SK, Nordin R, Ismail M, Ali JA. Accurate wireless sensor localization technique based on hybrid PSO-ANN algorithm for indoor and outdoor track cycling. IEEE Sensors Journal. 2016;**16**:529-541

[13] Swain RR, Khilar PM, Dash T. Neural network based automated detection of link failures in wireless sensor networks and extension to a study on the detection of disjoint nodes. Journal of Ambient Intelligence and Humanized Computing. 2018;**10**: 593-610

[14] Yuan Y, Li S, Zhang X, Sun J. A Comparative Analysis of SVM, Naive Bayes and GBDT for Data Faults Detection in WSNs. In: 2018 IEEE

International Conference on Software Quality, Reliability and Security Companion (QRS-C). Lisbon, Portugal: IEEE; 2018. pp. 394-399. DOI: 10.1109/QRS-C.2018.00075

[15] Cheng Y, Liu Q, Wang J, Wan S, Umer T. Distributed fault detection for wireless sensor networks based on support vector regression. Wireless Communications and Mobile Computing. 2018;**2018**:8. DOI: 10.1155/2018/4349795

[16] Abdullah MA, Alsolami BM, Alyahya HM, Alotibi MH. Intrusion detection of DoS attacks in WSNs using classification techniuqes. Journal of Fundamental and Applied Sciences. 2018;**10**:298-303

[17] Zhang D, Qian L, Mao B, Huang C, Huang B, Si Y. A data-driven Design for Fault Detection of wind turbines using random forests and XGboost. IEEE Access. 2018;**6**:21020-21031

[18] Zhang X, Zou J, He K, Sun J. Accelerating very deep convolutional networks for classification and detection. IEEE Transactions on Pattern Analysis and Machine Intelligence. 2016;**38**: 1943-1955

Chapter 5

Wireless Sensor Networks Challenges and Solutions

Sumana Naskar

Abstract

In this field, many research works have been done. And there is still emerging and trending domain for all researchers. Sub domains are also interested like WBAN, WPAN etc. as well as advance domains of this WSN is also interested and trending topic as IoT, IoE etc. In this fifth generation, IoT is everywhere. So, researchers can easily research on this topic and publish paper. Although advancement of this domain have been trending, but core WSN has many issues as disadvantages. Such as, it has seven layers. Each and every layer has many issues as challenges like simulator tools are very old configuration and simulator softwares are becoming obsolete. Hardware issues are there like devices should be supported with computer. Another important issues are sensor battery, energy consumption etc. Rather than those issues, most serious issue is security, because the lack of privacy constraints.

Keywords: WSN, WBAN, IOT, network security, energy consumption

1. Introduction

A sensor based network system that performs wirelessly has given birth to the coinage 'Wireless Sensor Network' or simply 'WSN'. 'WSN' [1] has become essential in many fields including infrastructure, healthcare, agriculture, environment, military defense, automation and so on.

As technological advancement is gaining its momentum day by day, it has been trying to replicate some of the biological receptive and cognitive faculties in order to develop more intelligent devices and systems to overcome real world hindrances. Thus 'sensors' [2] are used to imitate human sense organs to collect information about certain physical parameters like temperature [3], light, speed, movement, moisture etc. for any desired action taken thereafter.

Wireless Sensor Network is a system for the transmission of data through such a networking process which is independent of any physical cables. Here the sensors deployed receive some physical quantities like temperature, pressure, sound, humidity etc. from the environment and process them. They not only monitor the sense but also have the communication capability. Thus we can define Sensor as a device which measures any physical quantity and converts it to such a format which an observer or an instrument can understand (**Figure 1**).

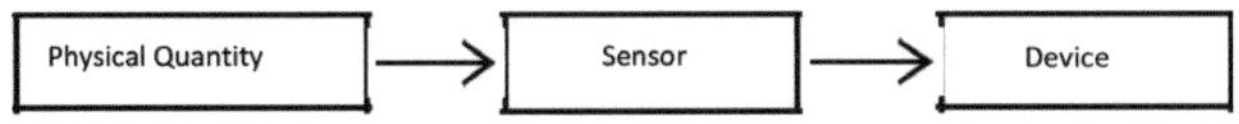

Figure 1.
Brief structure of WSN.

2. Architecture of WSN

A wireless sensor network [4–7] consists of Base station and many sensor nodes. The primary constituent of a Wireless Sensor Network is a cluster of interconnected sensor nodes. Wireless sensors are highly distributed, lightweight nodes deployed in large number to monitor the environment or system. These sensors constitute a network which is distributed on ad-hoc basis, which means they work on wireless connectivity and sensor nodes are fitted with on board processor.

These sensor nodes can have many features namely Sensor Subsystem, Processing system and Communication System. Now we are going to discuss these features briefly.

1. Sensor Sub-system: This is a combination of proprioception, visual, audio, inertial and tactile units. Proprioceptors are the sensory receptors which receives stimuli from within the body, especially one that responds to position and movement. Visual unit comprises high speed and high definition cameras. Similarly an audio system equipped with microphones, inertial measurement unit, force sensitive resistors are used to construct the sensor subsystem.

2. Processing system: The processing unit comprises Microcontroller unit (MCU), Memory and an operating system. The MCU performs computations on the received data, computes the next hop to the sink, controls and monitor battery power etc. In the Memory part program code is stored and it computes the procedure by nonvolatile RAM. Sensor uses operating systems like Contiki, TinyOS etc.

3. Communication system: Here data are exchanged between the sensors through radio transceiver. This unit receives query and transmits the processed data from MCU to the devices.

4. Power generator – It provides energy for sensor node and has two parts within namely Battery and DC-DC converter.

Optional components –

5. Location finding system: This system comes into use if the user requires the knowledge of location with high accuracy.

6. Mobilizer: It is needed to move sensors when it is required to carry out the assigned tasks based on its decisions on its mission depending on the routing algorithms used.

On the other hand Base Station is an end point device that gets all the processed data from the sensor nodes and it performs aggregation as well as fusion for the purpose of analysis and controlling it.

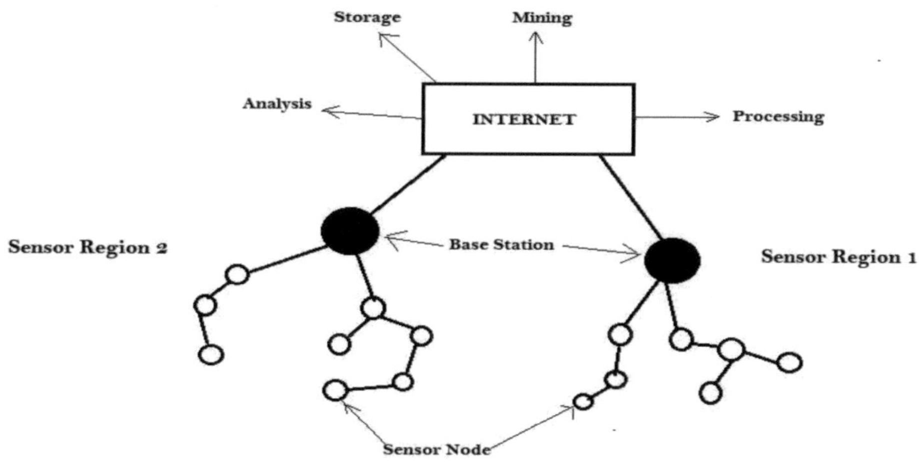

Figure 2.
Architecture of WSN.

Now we will see how a wireless sensor network works. A number of interconnected sensor nodes are arranged and rearranged to make a group and it is connected to a Base Station. As mentioned earlier, Base Station acts as the terminal point of a sensor network system and collects all the information from the group of sensor nodes. This Base Station is eventually linked to the internet. There could be other sensor node- Base Station systems connected to the internet. Thus a complex network of sensors are formed to get various data communicated through the internet so that the information can be used for different purpose according to the user's need like processing, mining, storing or analysis (**Figure 2**).

3. Different kinds of wireless network

Network means a group of devices interconnected to perform specified task. To achieve that purpose we have two distinct technologies namely 'wired network' and 'wireless network'. Between these two, 'wireless network technology' is often preferred in terms of flexibility, user mobility, network access and so on. In 'wireless sensor network' data is transferred through radio frequency unlike a wired network where data transmission takes place through Physical cables. There are different types of wireless network system in vogue depending on the data rate or coverage area (**Figure 3**).

W-WAN (Wireless Wide Area Network):-

As the name suggests, this is a network system that can provide long distance transmission of data, voice, image and video information over a large geographical area. The range of W-WAN is beyond 100 km. This kind of network is used in GSM, telephonic calls, surfing web pages etc.

Technologies used in W-WAN:

1. ISDN (Integrated Service Digital Network) – It is one of the technologies that is used to transmit data, voice and signaling and it is a circuit-switched telephone network system. It is well known for providing better speed and quality than traditional connections.

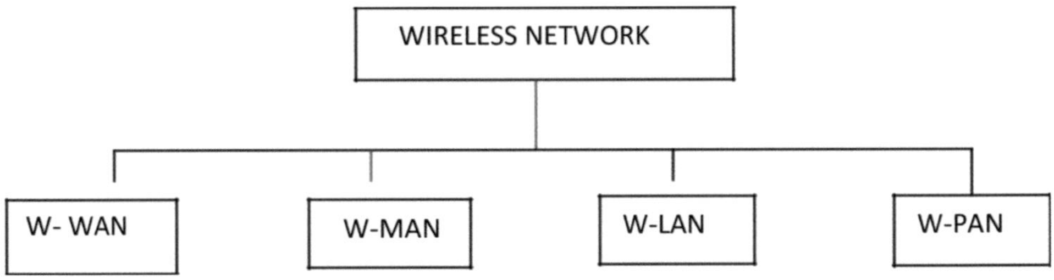

Figure 3.
Different kinds of WSN.

2. SMDS (Switched Multimegabit Data Service) – It is a packet-switched telecommunication service and is designed for high speed broadband networking technology.

3. SONET (Synchronous Optical Network) – It is a communication protocol used to transmit a great amount of data over large distances.

4. HDLC (High Data Link Control) – This is a data link protocol and data is organized into frames those are transmitted through the network to the destination that verifies its successful arrival.

5. SDLC (Synchronous Data Link Control) – An error-free movement of data is possible between the Network Addressable Units within a given communication network through this protocol.

W-MAN (Wireless Metropolitan Area Network):

This is a wireless network system which has the capability to cover an area of approximately the size of a city and it sits somewhere between W-LAN and W-WAN. The connections can be point to point or point to multiple networks.

W-LAN (Wireless Local Area Network):

A wireless LAN is a wireless computer network two or more devices using wireless connection to form a local area network within a limited area such as home, school, computer laboratory, campus or office building.

W-PAN (Wireless Personal Area Network):

A personal area network is a computer network for interconnecting electronic devices within an individual person's workplace. A PAN provides data transmission among devices such as computers, smartphones, tablets [8] and personal digital assistants.

4. Application of wireless sensor network

With the advent of new technology like WSN a plethora of areas in modern civilization has been redefined with newer prospects. WSN is introduced in the following applications:

i. Battlefield – Surveillance and monitoring the movement of enemies and own military forces.

ii. Disaster relief operation – receiving data and analyzing situation of the affected area.

iii. Environmental use – monitoring different environmental parameters of a particular region like temperature [9], air pressure, rain etc.; tracking related data of flora and fauna of certain biosphere.

iv. Agriculture – monitoring data regarding soil, weather, irrigation for agricultural use.

v. Health care [10]– monitoring patients' physical condition and giving necessary feedback in alarming situation.

vi. IoT [11, 12]– The Internet of Things works on the basis of physical world of devices and objects connected over the network using the wireless sensors.

4.1 Challenges of WSN

In spite of their highly practical usefulness there are some challenges [13] in wireless sensor network system –.

i. Scalability – There are a vast difference in scale of such sensor networking system as the number of sensor nodes may vary from few to several. Added to this the deployment density is correspondingly adjustable.

ii. Energy efficiency [14, 15]– As wireless sensor nodes have to work on a limited power supply, the designing of the software and the hardware has to be so optimized that it can perform efficiently the designated job.

iii. Maintenance [16, 17] – WSN has several constraints like power supply, storage, large amount of algorithms, so there is a serious challenge in maintenance of all these.

iv. Security [18] – Like all internet dependent applications, WSN also has insecurity scare. Proper data transmission management should be adopted to counter data theft by every possible way.

v. Quality of service – The data must be provided in time as the real time based applications heavily dependent on the timely distributed data.

4.2 Advantage of WSN

It is scalable and can accommodate any new node or device at any home. It is flexible and open to physical partitions.

- All the WSN nodes can be accessed through a centralized monitoring system.

- WSN can be applied on large scale and in various domains such as mines, agriculture, health care and so on.

- It uses different security algorithms as per wireless technologies and hence provide reliable network for customer user.

Disadvantage of WSN:

- It cannot be used in high speed communication as it is designed for low speed application.
- It is quite expensive to build such network, so cost-effectiveness of such system may be a concern for some users.
- WSN has limited computation and communication resources. It is prone to security threat.

5. Research issues on WSN

In this field, many research works have been done. And there is still emerging and trending domain for all researchers. Sub domains are also interested like WBAN, WPAN etc. as well as advance domains of this WSN is also interested and trending topic as IoT, IoE etc. In this fifth generation, IoT is everywhere. So, researchers can easily research on this topic and publish paper. Although advancement of this domain have been trending, but core WSN has many issues as disadvantages. Such as, it has seven layers. Each and every layer has many issues as challenges like simulator tools are very old configuration and simulator softwares are becoming obsolete. Hardware issues are there like devices should be supported with computer. Another important issues are sensor battery, energy consumption etc. Rather than those issues, most serious issue is security, because the lack of privacy constraints. Some main challenges are followed as:

1. Node deployment:- In WSN, sensor node deployment is the main challenge. Deployment should be always in proper way. If useless deployment is there i.e. in any vital area, there is needed more sensor to collect data, but rather that important deployment, remote area is getting more sensor which is totally useless.

2. Selection of Relay node:-Sometimes, sensor data is sent from source to destination (sink/base station) for collection of information. To get shortest path, sometimes source nodes information are sent through via node(relay node) to destination. To choose this relay node and shortest path, various methods must be followed. This is very hectic process to choose which method is best for this purpose. And this is very time consuming.

3. Selection of cluster head:- Like relay node, cluster head selection is very challenging. Because, some area has grouping node deployment. In that case, Cluster nodes, cluster node size, cluster head should be selected to send data to coordinator (base station). Basis on this selection, routing protocol has been designed.

4. Energy consumption:- WSN is based on sensor node. So, if sensor node battery is dead, communication will be disrupted. Therefore, battery power is needed to save for longtime rather consumes more energy.

DOI: http://dx.doi.org/10.5772/intechopen.109238

5. Security issues:- In this domain, this is a big issue. Anytime any data can be lost, hacked, misled via attacker. Malicious nodes, corrupted messages are also treated as security issues.

6. Heterogeneity:- Sometimes this is very much confused, at where in which node structures are needed. For health monitoring purpose, heterogeneous nodes are preferable.

7. Pathloss:- To get multipath for sending data in shortest way, many problems have been faced. Most of those, pathloss is very crucial problem. This is not only happened with multipath but also happened with single/direct path.

8. Delay:- During data transmission, messages communication, time is very critical measurement. Because, a nanosecond delay in WSN means a lot. Message deliver, acknowledgement can be failed due to this delay.

9. System failure:- Primary reason behind this is hardware-software failure, physically damage sensor node etc. Both client server architecture in WSN may face this type of situation.

6. Effective solutions

Although this field has so many issues, but still researchers are trying to solve those problems doing researches on it. Advance Simulator with compatible hardware, devices are being upgraded by computer engineers, so this will be helpful to research work and researchers do many update researches on WSN. More or less, researches are trying to put idea about network lifetime of sensor battery. Also, they are making cryptography algorithm to provide more security about field. Some solutions are given as followed below:

1. Node deployment structure should be maintained non-adhoc network structure i.e. well organized way and well planned protocol manner.

2. Relay node selection should be based on Euclidean distance manner i.e. equal distance from relay to source node should be followed.

3. Cluster head node selection should be based on k-means clustering, KNN algorithm. Other better algorithm can be fitted.

4. Energy consumption can be less and it can be happened due to following multi hop routing protocol. If shortest path is chosen, then energy can be less consumed. It is direct proportional to network lifetime. So more energy saving means more longevity of sensor battery node and as well as total network longevity.

5. To give security, data confidential, data integrity, data authentication concept is there. To stop stealing data, data should be kept in confidential way using cryptography techniques. Some mechanisms such as verification should be implemented to check for authenticated data. Data integrity always check receiving data is matched with previous sending data or not.

6. Fault tolerance is the technique to tolerate all types of failure within WSN, it measures failure of the system.

Author details

Sumana Naskar
Prajnananda Institute of Technology and Management, Kolkata, India

*Address all correspondence to: sumananaskar1@gmail.com

References

[1] Ko JG et al. Wireless sensor networks for healthcare. Proceedings of the IEEE. 2010;**98**(11):1947-1960

[2] Taleb T et al. ANGELAH: A framework for assisting elders at home. IEEE Journal on Selected Areas in Communications. 2009;**27**(4):480-494

[3] Kelly SDT, Suryadevara NK, Mukhopadhyay SC. Towards the implementation of IoT for environmental condition monitoring in homes. IEEE Sensors Journal. 2013;**13**(10):3846-3853

[4] Ullah S et al. A comprehensive survey of wireless body area networks. Journal of Medical Systems. 2012;**36**(3):1065-1094

[5] Liang X et al. Enable pervasive healthcare through continuous remote health monitoring. IEEE Wireless Communications. 2012;**19**(6):10-18

[6] Yuce MR. Implementation of wireless body area networks for healthcare systems. Sensors and Actuators A: Physical. 2010;**162**(1):116-129

[7] Mitra U et al. KNOWME: A case study in wireless body area sensor network design. IEEE Communications Magazine. 2012;**50**(5):116-125

[8] Morak J et al. Design and evaluation of a telemonitoring concept based on NFC-enabled mobile phones and sensor devices. IEEE Transactions on Information Technology in Biomedicine. 2012;**16**(1):17-23

[9] Min AW, Zhang X, Shin KG. Spatio-temporal fusion for small scale primary detection in cognitive radio networks. In: Proc. IEEE INFOCOM. 2010. pp. 1-5

[10] Lu K, Qian Y, Chen HH, Fu S. WiMAX networks: From access to service platform. IEEE Network. 2008;**22**(3):38-45

[11] Nasr A et al. Cloud-based virtualization environment for IoTbased WSN: Solutions, approaches and challenges. Journal of Ambient Intelligence and Humanized Computing. 2022;**13**(10):4681-4703

[12] Hugo L et al. A review of IoT sensing applications and challenges using RFID and wireless sensor networks. Sensors. 2020;**20**(9):2495

[13] Salman ID et al. Challenges and issues for wireless sensor networks: A survey. Journal of Global Scientific Research. 2021;**6**(1):1079-1097

[14] Amendola S et al. RFID technology for IoT-based personal healthcare in smart spaces. IEEE Internet of Things Journal. 2014;**1**(2):144-152

[15] Montgomery K et al. Lifeguard—A personal physiological monitor for extreme environments. In: Proc. 26th Annu. Int. Conf. IEEE Eng. Med. Biol. Soc (EMBS). 2004. pp. 2192-2195

[16] Shnayder V et al. Sensor networks for medical care. In: Proc. 3rd Int. Conf. Embedded Netw. Sens. Syst. ACM (SenSys). 2005. pp. 313-314

[17] Wood AD et al. Context-aware wireless sensor networks for assisted living and residential monitoring. IEEE Network. 2008;**22**(4):26-33

[18] Rami A et al. Machine learning for wireless sensor networks security: An overview of challenges and issues. Sensors. 2022;**22**(13):4730

Chapter 6

Novel PTC Composites for Temperature Sensors (and Related Applications)

Radu Setnescu and Eduard-Marius Lungulescu

Abstract

This chapter presents a brief description of conductive polymer composites in general, with more attention paid to those exhibiting abrupt change of resistivity when temperature raises, associated with PTC and NTC (respectively, positive and negative temperature coefficient of resistivity) effects. These materials are "smart" because they can adapt their electrical behavior to environmental characteristics, hence being suitable for temperature sensors, smart heating devices, safe batteries, and resettable fuses. As compared to NTC, the applications of PTC materials are more numerous, because the abrupt increase of resistivity with temperature rise allows the current and temperature to be naturally limited. The PTC effect and the factors controlling its quality, e.g., repeatability, intensity, switching temperature, and subsequent NTC effect, are discussed from the point of view of the influence of the nature of polymeric matrix, conductive fillers, and applied treatments. Increased attention is paid to composites with conductive carbonaceous fillers, and these materials being of great interest because they have considerably lower density than metals, are easier to process, and can impart surprising mechanical and electrical properties to polymer matrices. Examples and applications of temperature sensors based on PTC composite materials, applications, and perspective aspects are discussed within the chapter.

Keywords: temperature sensors, thermistor, positive temperature coefficient, conductive polymer composites, carbonaceous materials

1. Introduction

A sensor is a device that responds to a chemical or physical stimulus by generating a signal that can be analyzed electronically. In general, to be suitable, sensors and devices incorporating them must have a fast response to external stimulus, be able to detect fine variations in the analyte, have a low recovery time, identify the desired analyte among other stimuli, and be a reliable tool, easy to operate, preferable also in *in situ* measurements [1]. In addition, it is desirable to have a low cost and environmental compatibility. Although there are a large number of sensors proposed in the literature, continuous concerns for the improvement of sensors are justified by the need to improve existing systems as well as to find new promising alternatives that allow the change of old analysis and measurement technologies. A widely investigated possibility in the last decades is the use of nanomaterials, a category that seems to offer an infinite practical field of solutions for the most diverse problems. Among the various nanomaterials, nanostructured carbonaceous materials show numerous advantages compared to other materials due to their specific properties as well as convenient costs, which make them suitable for use as technological sensors. Therefore, special attention will be paid in this chapter to recent advances in the field of resistive temperature sensors using such materials.

Temperature sensors are indispensable in various fields of engineering (electrical, automotive, aerospace, communications, civil engineering [2], health care, human machine interface, robotics, etc. [3]). Therefore, among the sensing technologies, the temperature sensor is probably the most widely used. The signal given by resistive temperature sensors consists of the increase or decrease of resistivity or capacitance with temperature [2]. The continuous improvement of the performance and manufacturing processes of sensors in general and of temperature in the present case, as well as the expansion of application fields, are current concerns of materials science and related disciplines.

Among the directions of perspective or current interest for thermal or multifunctional sensors that include temperature detection are flexible electronics [2] for intelligent house applications (e.g., electronic wallpaper, an interactive system that incorporates a network of sensor temperature and an air conditioning control system [4]), robotics (humanoid artificial skin [5, 6]), control of working conditions in mechanical or electronic systems, instruments wearables for physiological monitoring, smart packaging that indicates the state of freshness of food, etc. [6, 7]. In such applications, portable temperature sensors, flexible themselves, capable of detecting several signals simultaneously (temperature, pressure, and strain), the so-called multifunctional sensors (or physical sensors) are often required, or, on the contrary, the temperature signal should not be affected by the action of other factors, such as mechanical deformation (monofunctional sensors, in our case, temperature sensors) [7]. Some of the portable, flexible sensors proposed in recent years are PTC resistive sensors [6, 7].

A major obstacle to the practical realization of stretchable resistive temperature sensors from composite materials is the limited ability of the conductive phase to be stretchable during use, that is, to retain its sensitivity unchanged to repeated, large-amplitude deformations. Obviously, work must be done on the conductive phase, by finding conductive nanoparticles, capable of favorable interactions with the macromolecular chains of an optimally chosen elastomeric matrix. For this purpose, metal nanoparticles, carbon nanotubes (CNTs), graphene, as well as different preparation techniques, have been tested. The polymer substrates used for flexible printed sensors are mainly polymers containing aromatic rings in their structure, such as polyethylene terephthalate (PET), polyethylene naphthalate (PEN), and polyimide (PI), which are available as films with large surface areas, at reasonable prices, allowing the obtaining of reasonable production costs [8].

Stretchable (elastic) temperature sensors are usable in a wide range of applications, such as wearable real-time health care devices, monitoring working conditions in mechanical or electronic devices, wearable instruments for physiological monitoring, and smart packaging. Their realization involves the use of matrices or elastic supports. Temperature sensors often need to accompany stretchable strain sensors in applications such as wearables for recording human physical movements, therapy devices, and health monitoring such as heart rate monitoring [2]. Extensible strain and temperature sensors are also required for measuring deformations on extensible and curved surfaces (deformation of the skin on the wrist, e.g.) or for measuring the stress developed on the curved surfaces of pressure vessels, often requiring that local temperature information to accompany the strain information.

The properties of conductive polymer composites (CPCs) as temperature sensors have been reviewed in various previous works [9–12].

This chapter is dedicated to the recent progress made in the field of temperature sensors with PTC properties and their applications, which tries to develop and complement the previous information, aiming also to give an overview of the subject. Although not the subject of this chapter, some notable examples of NTC sensors are also included.

2. Conductive polymer composites. PTC effect

Unlike intrinsically conductive polymers (see [13–15], such as polyacetylene, polyaniline (PANI), polythiophene, polypyrrole (PPy), poly(p-phenylene vinylene), or PEDOT:PSS), the polymer composites, consisting of an electrically insulating polymer and a conductive phase, are much easier to prepare and have high chemical stability, and their electrical conductivity can be varied simply, within very wide limits, by changing the nature and the concentration of the conductive filler, changing the nature of the polymers used as a matrix, as well as by changing some technological factors (such as mixing the components, heat treatments, and/or irradiation, pressure). Common polymers are known to have intrinsically poor electrical and thermal conductivity, making them useful as insulating materials. The addition of a conductive filler to the polymer matrix gradually worsens the dielectric properties of the material, which at some point, quite suddenly, becomes conductive; the concentration value at which the insulator-conductor transition occurs is called the percolation threshold. Compounding polymers with conductive fillers has become a method to obtain a wide range of materials with different electrical properties, including those generically called conductive polymer composites. In recent years, there has also been a growing interest in making composite materials using intrinsically conductive polymers, as well as integrating them into a range of electronic devices, including sensors (see [16]).

2.1 Polymeric matrices

All categories of polymers—thermoplastics, elastomers, thermosets, synthetic polymers, or natural polymers are suitable in principle for making polymer matrices, as can be seen from the examples presented below. Elastomeric conductive composites (ECCs) based on carbon fillers are very attractive and play a significant role in the field of smart sensors due to their excellent flexibility, wide sensitivity spectrum, as well as fast response to external stimuli.

A current trend is also the growing interest in biodegradable matrices such as polylactic acid (PLA), giving to the products that incorporate them a lower environmental impact than synthetic polymers.

2.2 Fillers

Combining polymers with electrically and/or thermally conductive nanofillers allows the development of polymer nanocomposites in the form of thin, light, flexible, wearable films that present interesting electrical and thermal properties for various technical, environmental, and biomedical applications. As conductive fillers for obtaining composite films applicable as sensors in general, various structures are mentioned, as for example Au, Ag nanoparticles, Au or Ag nanowires [2, 8], Ni microparticles [17], and carbon-based materials (e.g., graphene or reduced graphene oxide (rGO) [18], CNT [19] and carbon black (CB)[10], and ceramics (e.g., $Mn_{1.71}Ni_{0.45}Co_{0.15}Cu_{0.45}Zn_{0.24}O_4$ [20]) often together with newer manufacturing technologies such as additive manufacturing (AM) through 3D printing [21].

The appearance of electrical conductivity at the percolation threshold can be explained by a simple intuitive model, in which it is considered that each filler particle is in physical contact with the neighboring ones, thus achieving electrical contact throughout the matrix. Another mechanism for achieving conduction, which explains reaching the percolation threshold at concentrations lower than those corresponding to the physical overlap of the particles, is that of tunneling; in this case, the distance between two filler particles must be max. 1.8 nm [2, 22].

For a given polymer, the electrical resistivity of the resulting composites can be varied over a very wide range by changing the nature and concentration of the filler, as well as the production technology [23]. As the concentration of conductive filler in the polymer matrix increases, the resistivity of the material gradually decreases, reaching a critical concentration (φc) at which the insulator/conductor transition is observed for the studied composite. The value *φc* represents the conductive percolation threshold, according to the classical theory of percolation. For concentrations of the conductive phase, φ > φc, the value of *φc* can be determined with relation (1) [24, 25]:

$$\sigma \propto (\varphi - \varphi_c)t \tag{1}$$

where φ is the electrical conductivity of the composite and t is an exponent that depends mainly on the size of the conductive network in the composite [10].

The decrease in resistivity and the appearance of conductive properties is determined by the formation of conductive networks inside the polymer, in which conductive particles are in electrical contact, allowing the passage of electric current from one conductive particle to another (through tunneling or jumping mechanisms). The maximum distance between particles in such a conductive path has been estimated to be several nanometers [26]. Not all conductive particles dispersed in the polymer participate in the transport of electrical charges through the sample (i.e., they are not effectively involved in conductive paths), but only a considerably smaller fraction than the filler fraction contained in the polymer (**Figure 1**). For example, in the case of CNT, it was estimated that approximately 3.3% of the particles are involved in the effective conduction of the current (being part of the so-called backbone of the electrical conduction (**Figure 2**), in fact, a conductive path similar to those in **Figure 1**, for spherical particles of CB).

The composite materials, also called hybrid materials, are biphasic or multiphase systems, which present special properties, resulting from the synergistic combination of components [1, 27]. Conductive polymer composites have, due to their remarkable properties (e.g., low density, high mechanical strength and corrosion resistance, and controllable electrical conductivity over a wide range, up to values close to the conductivity of metals), numerous applications (e.g., space constructions, automotive industry, sporting goods, electronic devices, electromagnetic shielding, energy storage (safe batteries), overcurrent protections (resettable fuses), filters, and sensors [28])

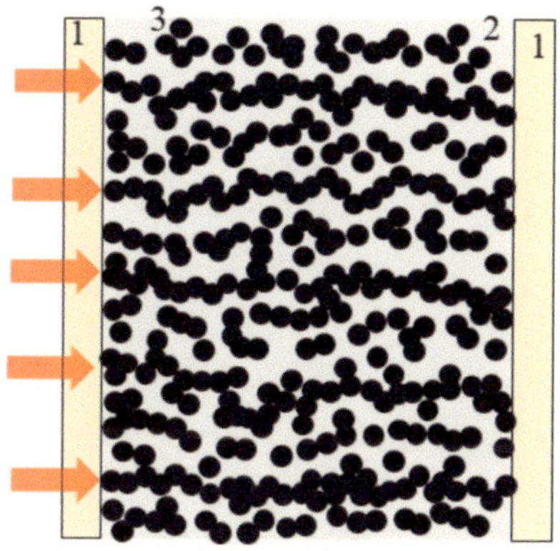

Figure 1.
Possible distribution of small spherical particles (e.g., CB) within a polymer composite: 1—electrodes; 2—polymer matrix: 3—conductive particles (CB). The red arrows indicate the conductive paths which enable the electrons to across the sample. $\vec{\mathbf{E}}$ = *electric field used to reveal the conductivity.*

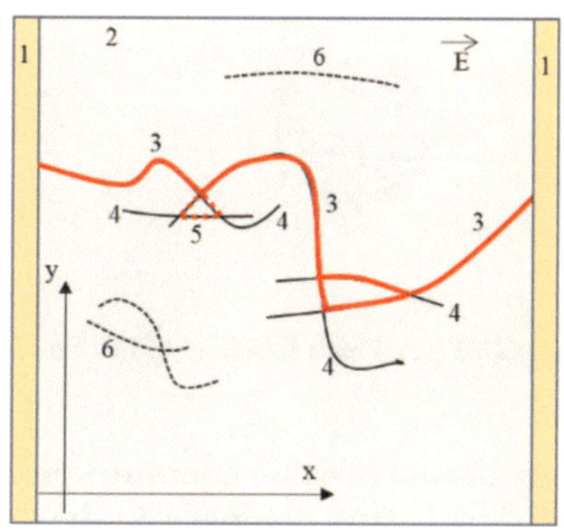

Figure 2.
Possible CNT distribution within a 2D CNT/polymer composite (e.g., a thin film sample): 1—electrodes; 2—polymer matrix; 3—conductive path (backbone of conductive network); 4—zero-current branches; 5—balanced branches; 6—isolated CNT clusters; $\vec{\mathbf{E}}$ = *electric field used to reveal the conductivity (Adaptation after [26]).*

where they have successfully replaced "traditional" materials, such as ceramics or metals [10].

The influence of the filler type, polymer matrices, and dispersion methods on the electrical properties of the resulting CPCs has been reviewed by different authors (see [29] for segregated CPCs, [11] for control methods of electrical properties, and [30] for correlating CPC electrical properties with phase morphologies such as phase segregation or co-continuity [9]).

An effective way to reduce the conductive filler content (or percolation threshold) is to make segregated composites (s-CPCs). The polymer matrix can consist of two thermoplastic, immiscible polymers, e.g., polystyrene (PS) and polypropylene (PP) [31], with the conductive filler (e.g., a carbonaceous material) being preferentially distributed in one of the polymers (PS), and the condition φ_c for the respective polymer is achieved (**Figure 3**). The conductive paths are thus formed at the interface of the two polymers, the composite as a whole becoming conductive at very low concentrations of the conductive filler, compared to homogeneous composites [29]. Other possibilities for obtaining segregated CPCs are shown in **Figure 3**. Comparing

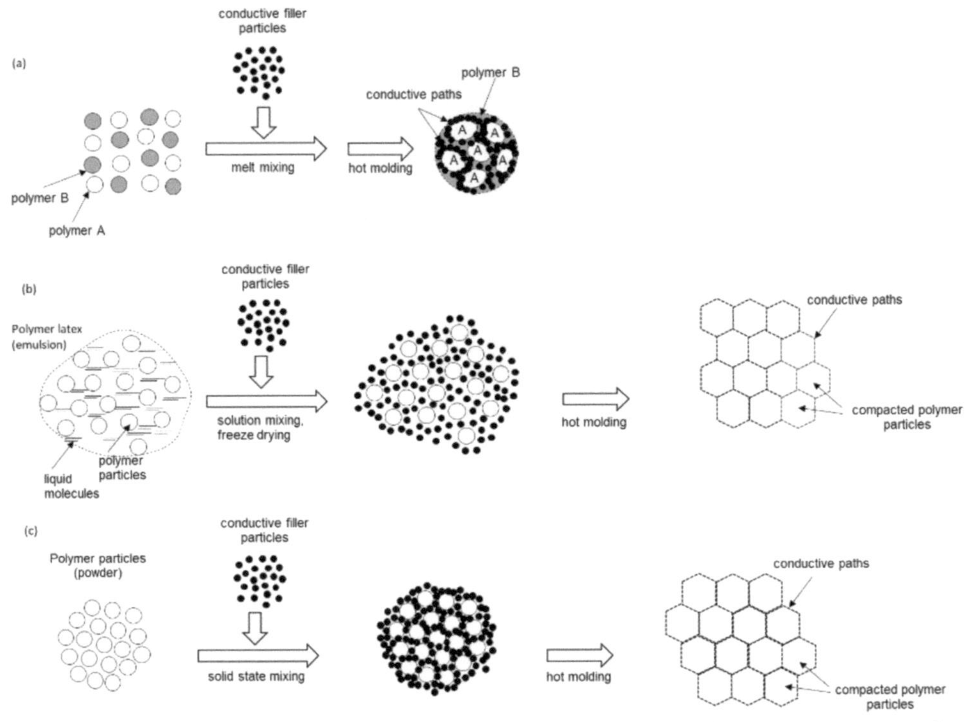

Figure 3.
Three simple ways to produce segregated CPCs: a) melt blending; b) latex technique; and c) dry or solvent mixing (adapted from [29]).

Figures 1 and **3**, it intuitively results that the concentration required to create a network of conductive paths is substantially lower in the case of s-CPC.

Ideally, carbonaceous materials have a structure consisting exclusively of carbon, which can be found in various states of sp^n hybridization, n = 1–3. We must also take into account the fact that, often, there may be heteroatoms (mainly hydrogen or other, oxygen, sulfur, and nitrogen), especially at the edges of extended carbon domains. In general, carbonaceous materials can be characterized by the state of hybridization, as well as by the ratio of the number of hydrogen to carbon atoms (H/C) [32]. Among the carbonaceous materials, those that have potential for use in composites are those that present electrical conductivity. For the sp^3 hybridization state, the typical representative, diamond, does not conduct electricity because all the valence shell electrons move in well-defined directions along the σ bonds. In the case of the other types of hybridization, sp. and sp^2, the electrons in the unhybridized *p*-orbitals have greater freedom of movement and, in the case of extended conjugation effects, can cause the electrical conductivity of the respective substances to appear. It follows that the appearance of electrical conduction will occur in the case of large molecules (polymers or oligomers), in which there may be a large number of conjugated bonds, leading to the formation of molecular orbitals extended on a molecular scale, in which *p* electrons can move practically free.

Sp-hybridized compounds that fall under this condition are polyynes (or carbynes), which formally originate from the polymerization of acetylene and are a linear allotropic form of carbon. However, these compounds currently have a rather theoretical importance, not existing in nature, and their synthesis and stability remain quite unclear. The only example of synthetic carbyne comes from synthesis in carbon nanotubes, when chains of several thousand sp. carbon atoms were linked to form polydisperse carbyne samples [33–35]. Among the attractive physical properties, the carbine is the unusual electrical transport of electrical charges (negative differential resistance [35, 36]).

Electrical conduction in carbonaceous materials has been studied by different authors and reviewed in a number of papers, such as those by [37, 38]. It is interesting to note that CNTs and graphene nanoribbons show ballistic conduction; that is, the transport of free electrons takes place over relatively long distances (longer than the active length of the medium in which the displacement of charge carriers occurs). Thus, these materials show practically zero resistivity, although they are not superconductors.

Carbon materials with current practical uses involving electrical conduction have sp^2 hybridization, exhibit continuous conjugation, and can be classified as follows [32]:

- Graphene (Gn) and graphite (Gr), which represent an infinite plane of sp^2-hybridized C atoms and respectively a vertical overlap of such planes that interact with each other through weak van der Waals forces; the p electrons from orbitals not participating in the hybridization are contained in an orbital extended to the scale of the entire graphene plane, their delocalization determining the electrical conduction along the graphene plane as well as the anisotropy of the electrical properties of graphite;

- Graphene fragments (which include nanocarbon and nanographene), which have H/C atomic ratio values between 1 and 0, usually between 0.5 and 1, depending on the degree of development of the carbon skeleton;

- Fullerenes (FLN) and CNT do not contain structural hydrogen, but their hybridization state is slightly different from sp^2, because the morphology of the respective molecules presents curved surfaces; therefore, the hybridization state must be between sp^2 and sp^3 (or to put it another way, the π conjugation is also robust on curved surfaces). Fullerenes are an allotropic form of carbon whose molecules are in the form of closed cages (Cn) formed by rings with 5 or 6 carbon atoms; in each fullerene structure, there are 12 such 5-carbon rings and a variable number, depending on the size of the fullerene, of 6-carbon rings [1].

- Carbon nanotubes structurally belong to the same family as fullerenes. Single-walled (SWCNT) or multi-walled (MWCNT) structures are known, so CNTs can be classified according to the number of graphene layers in their structure. A SWCNT is a cylindrical tube with a diameter of 0.5–1 nm covered by hemispherical ends, which thus appears as a roll, being equivalent to a rolled graphene sheet [39]. MWCNTs comprise several such concentric cylinders, having a diameter of 2–100 nm and a distance between layers of 0.3–0.4 nm. Since the distance between graphite interlayers is similar, MWCNT can be viewed as a folded graphite sheet [40].

- the group of materials generically called amorphous carbon that contains coal in its various forms (coal, brazier, charcoal, etc.), soot, carbon black, etc. These materials are important for industry in general and contain, in addition to C and hydrogen, other heteroatoms such as O, N, or S; taking into account the composition, carbon fibers can also be included in this group, although structurally, they are more similar to graphite, and the precursor of vapor-grown carbon fiber (VGCF) is very similar to CNT [32].

Graphene, a monolayer (atomic single crystal) structure of covalently bonded sp^2-hybridized carbon atoms, is currently the thinnest material known in the world. Due to its structure, graphene exhibits a series of interesting properties, distinguished by their values, compared to common materials [2, 41], such as: high mobility of charge carriers (electrical conductivity can reach 6000 S/cm, comparable to CNT), high thermal stability (~2600 K [42]); very good thermal conductivity (5300 W/(m·K) [42]); special mechanical properties: mechanical modulus (~ 1 TPa) and breaking strength (~ 125 GPa); gas barrier properties, high transparency, and high-specific surface area [41].

Therefore, graphene has great application prospects and market value in various fields such as electric current transport, high-frequency electronic devices, flexible display, sensors and biosensors, batteries, supercapacitors, aerospace, and biomedical technologies. Graphene is also an ideal nanofiller for reinforcing polymers (composites). Even a small amount of graphene added to the composite causes a spectacular increase in mechanical, electrical, and processing properties [41, 43].

The various known methods for obtaining (synthesizing) graphene are reviewed in the literature (see [43]). Since graphene is difficult to produce (requires a lot of energy and is difficult to structurally control), being consequently expensive, different forms of modified graphene are used in practice (**Figure 4**), such as graphene oxide (GO) and reduced graphene oxide (rGO), as well as graphene wafers, which represent more advantageous alternatives in terms of production costs (see [43, 44] for synthesis of GO and subsequent rGO). Like graphene, these materials can induce great enhancement when combined with polymers, resulting in low-density, corrosion-resistant, low-friction, and low-cost nanocomposites with high-performance and multifunctional properties. However, since the oxygenated groups of GO can significantly affect the properties of the graphene structure, the functionalization of GO is often practiced, whereby the influence of the oxygenated groups is blocked by physical bonds (noncovalent functionalization) or by chemical bonding (covalent functionalization); see more details within the review [43].

Graphene platelets (GnPs) consist of several graphene layers (~3) overlapped. The assembly has an average thickness of 3.55 ± 0.32 nm [2, 45], thus being considered a 2D nanomaterial [46]. It exhibits high mechanical strength and high electrical conductivity (~1400 S/cm), as well as good compatibility with most polymers, as well as preparation of cost efficiency [46]. An interpretation of Raman spectroscopy data in

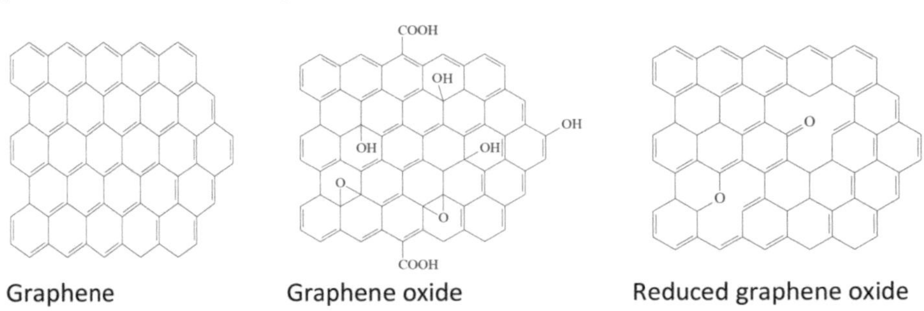

Figure 4.
Graphene, graphene oxide, and reduced graphene oxide.

terms of confirming the structure and assessing its integrity [2], compared to GO and rGO, the use of graphene platelets is more efficient in terms of the costs associated with the preparation, as well as the lower concentration of defects compared to GO [45].

A simple method to prepare graphene platelets consists of a thermal shock applied to a graphitic intercalation compound (GIC), followed by an ultrasound treatment [2]. A more detailed description is given in the reference [45].

Polymer/graphene nanocomposites exhibit suitable properties for applications as sensors, and many other electronic and mechanical devices such as capacitors, electromagnetic shielding systems, transistors, electroluminescent devices, batteries, memory, gate dielectric devices, light-emitting diodes, devices with touch screen, and solar cells [41]. A recent review of applied methods for obtaining polymer composites with graphene and modified graphene is presented in the reference [43].

Since the first investigations on the PTC effect in MWCNT/high density polyethylene (HDPE) composites, it has been observed that the presence of CNTs in the polymer matrix can considerably improve the thermal stability of this type of composites due to the interpenetration of the molecular chains of CNTs and HDPE [47]. Carbon nanotubes (CNTs) are thus ideal for inducing electrical conductivity in insulating polymers [26].

The use of CNTs in HDPE matrix composites led to thermistors with significantly better properties (in terms of holding voltage, current through the sample, and response time to applied voltage) compared to commercial CB-based thermistors, with potential critical applications of high temperatures, intense currents, and high applied voltages [48]. However, a problem that arises in the case of using CNT as a conductive filler is the agglomeration of CNT particles, caused by the attraction through van der Waals forces, which results in a reduced value of the weight of CNT particles that effectively participate in electrical conduction (see §2). An effective way to counteract this effect is to modify the surface of the CNT particles. For example, Zhou & Lubineau [49] modified MWCNTs by coating them with a conductive polymer, poly(3,4-ethylenedioxythiophene)poly(styrenesulfonate). After dispersion in a polycarbonate matrix, the electrical resistivity decreased by 11 orders of magnitude for a concentration of 1% (wt), compared to a decrease of only 8 orders for neat MWCNTs at the same loading.

Carbon black (CB) is the most used carbon material for making commercial CPCs, due to its low price, high availability, easy handling, as well as the good electrical conductivity imparted to the polymeric materials in which it is incorporated. However, a number of disadvantages have been identified that limit the exclusive use of CB in high-precision applications such as thermistors, as follows [48]:

- difficult processability of thermoplastic CPCs at high concentrations of CB (>25% wt), specific for obtaining low resistivities ($\sim$ 10 ohm);
- low thermal stability, determined by the susceptibility of CB particles to oxidation at high temperatures;
- poor thermal reproducibility, determined by the random recovery of conductive paths in thermal cycles.

To overcome some of these drawbacks, different solutions have been proposed to improve the properties of CPC with CB, such as: (i) radio-induced cross-linking, for the stabilization of conductive paths [24]; (ii) the use of mixtures of fillers with a synergistic effect on electrical conductivity [50], in order to reduce the content of conductive charge; (iii) the use of binary, heterogeneous, or homogeneous polymer matrices, in order to obtain the percolation effect at low concentrations of conductive charge or, respectively, to increase electrical reproducibility and improve machinability and mechanical properties; (iv) the use of fillers of a different nature, such as graphite, graphene, or CNT, whose particles having a high aspect ratio and greater electrical conductivity and thermal stability than carbon black can ensure the easier formation and stabilization of conductive paths [48].

The general interest in using carbonaceous materials as conductive fillers in sensor construction is due to their special properties, such as high specific surface area, chemical and mechanical stability, adaptability and functionality, and in the case of resistive sensors, electrical conductivity close to that of metals, as well as good processability with different polymer matrices [10]. Although metal powders are intrinsically more conductive than some carbonaceous materials, such as carbon black, the latter (as well as carbonaceous materials in general) is much more used to obtain conductive composites due to its high chemical inertness. It is known that metal particles tend to undergo oxidation, covering themselves with insulating films of oxides, which results in a decrease in the electrical conductivity of the composite over time [25]. In addition, polymer composite films and sensors with carbonaceous materials (especially those with CNTs and graphene) have promising light weight, flexibility, elasticity, sensitivity, and durability compared to their counterparts prepared with metal nanoparticles. The structure-property relationships of polymer/ graphene and CNT nanocomposites and films have been analyzed in a number of previous works [22, 46, 51].

2.3 The effect of temperature on the resistivity of CPCs

Based on the temperature dependence of resistivity, CPC materials can fall into one of the following three categories [52]: (i) PTC—positive temperature coefficient (resistivity increases with temperature); (ii) NTC—negative temperature coefficient (resistivity decreases with temperature); and (iii) ZTC—zero temperature coefficient (resistivity does not depend on temperature). From the multitude of known CPC materials, this chapter deals only with those applicable as resistive temperature sensors with PTC effect (thermistors); more tangentially, certain aspects of NTC sensors will also be covered. Obviously, ZTC materials are not interesting for resistive sensors because their resistivity is independent of temperature.

It is worth to mention that many of the CPC materials that exhibit PTC effect at temperatures lower than the transition temperature may exhibit NTC effect at temperatures higher than the transition temperature [50]. Although PTC materials are often used as self-temperature-regulating heating elements or protective devices with self-limiting current (resettable fuses), micro-switch sensors [52], in reality, in all these applications the respective materials functioning as temperature sensors/ thermistors, causing the temporary interruption of the passage of electric current through a certain circuit, when a certain designed temperature is reached, or maintaining an element at a constant, specific temperature. Similar materials with PTC effect can also work as thermometers/temperature sensors, based on a calibration curve, taking into account that their resistance increases with temperature, unambiguously (**Figure 5**).

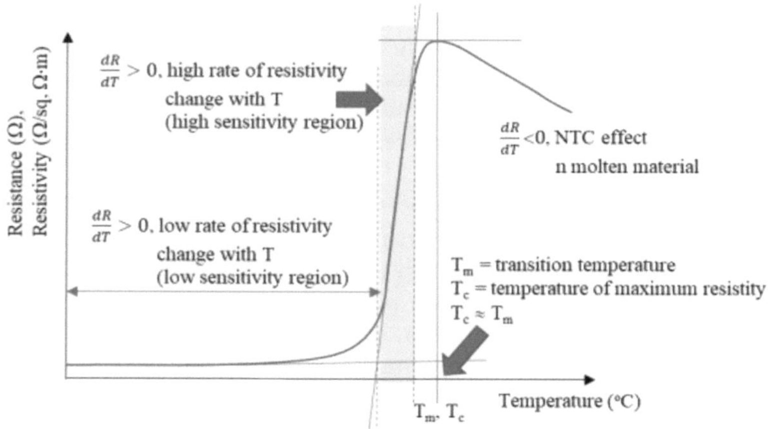

Figure 5.
Resistivity change with temperature for a composite showing PTC effect in solid.

PTC polymer composites have been actively studied due to their high performance and low manufacturing cost. A large amount of research work was initially carried out on PTC materials composed of semicrystalline polyolefins and CB filler, due to the high availability of these materials, easy processability, as well as low resistivity at room temperature [53].

3. Resistive temperature sensors based on CPC materials

Temperature sensors based on the change of some electrical properties can be of different types, such as *pyroelectric* (some ceramic materials or PVdF can generate a temporary voltage when the temperature changes due to the thermoinduced change of an internal polarization state), *RTD* (resistive temperature detector—change in the electrical resistance of a metal with temperature), or *thermistors* (see below) [6]; thermocouples (junction of two dissimilar metals that provide a temperature-dependent electrical voltage), and these are widely used in current practical applications but they appear to be less suitable for miniaturization and nanotechnologies than those already mentioned.

In a classical definition, thermistors were described as devices made of semiconductor materials whose resistance varies with temperature according to an exponential law, within certain limits. Inorganic thermistors were initially made by pressing oxide powders, or their mixtures, followed by sintering at characteristic temperatures. In general, the early thermistors were bulk materials, with NTC effect and allowed the detection of temperature variations of the order of $5 \cdot 10^{-4}$ °C. The characteristic temperature for inorganic thermistors depends on their type and is usually in the range of −100 ... + 300 °C. More recent literature indicates the use of ceramics (e.g., $Ba_{0.5}Bi_{0.5}Fe_{0.9}Sn_{0.1}O_3$/alumina or $Ni_{0.8}Co_{0.2}Mn_2O_4$/alumina) deposited by screen printing processes resulting in thick film thermistors. This type of thermistors has a considerably smaller size than bulk thermistors, allowing integration in microelectrical circuits [54]. However, the current interest is especially related to organic or hybrid materials, which are more suitable for fine applications in advanced fields such as biomedical and robotics.

Newer materials are conductive polymer matrix composites (CPCs), where the filler can be a metal powder, a semiconducting ceramic, or a carbonaceous material [10, 50]. The operating principle of temperature sensors based on the thermoelectric behavior of CPCs is that, upon heating, the conductive paths of the CPCs change, consequently also changing the electrical resistance (or resistivity/conductivity) of the material [10].

The performance of polymer composites as resistive sensors is closely related to the conduction mechanisms of charge carriers in the conductive filler as such, as well as in the composite as a whole. In general, the key factors that determine the behavior of sensors based on composite materials are polymer properties, conductivity and structural characteristics of the nanofiller, dispersion quality, as well as processing conditions [2].

Since the electrical properties of CPCs depend on the state of the conductive paths in the CPC matrix, the slight change of this state under the action of an external stimulus, such as mechanical deformation, pressure, temperature, and the presence of some liquids (organic solvents), can lead to a significant (measurable and unequivocal) variation of an electrical output signal (resistance, conductivity, and current). Therefore, CPC materials can be designed as sensors for the detection and quantification of external stimuli of the type already mentioned [10].

3.1 Temperature sensors obtained by hot forming (with thermoplastic matrix)

3.1.1 Generalities

Polymer composites for thermistors are conductive materials consisting of a polymer matrix (a single polymer or a mixture of polymers) and a conductive filler (or a mixture of fillers), which give the material electrical conduction properties. Black carbon [10, 25, 50, 55, 56], graphite [50, 57], graphene platelets [28], carbon nanotubes [28], carbon fibers [58], as well as metal powders [59] are commonly used as fillers.

PTC effect thermistors made of polymer composite materials are resistors with positive temperature coefficient of resistivity, applicable as temperature sensors. Depending on how the resistance varies with temperature, two types of PTC thermistors are distinguished: linear and commutative [27]. Linear thermistors are usable as temperature sensors over a wide thermal range, being characterized by a slow, practically linear increase in resistivity with temperature. Switching PTC thermistors are characterized by a steep jump in resistivity at a characteristic temperature and are therefore used as high-sensitivity temperature sensors for narrow thermal ranges located in the vicinity of the characteristic temperature [27]. In applications as resettable fuses or heaters with thermal self-regulation, the respective materials are actually temperature sensors that ensure the functionality of the device as current/voltage protection or as a heating element with self-limiting power [50]. In this regard, the intensity and reproducibility of PTC effects are important factors for the application of temperature sensors [60]. The use of PTC materials as temperature sensors is based both on the slow increase of resistivity with temperature (on the low temperature portion of **Figure 5**) and on the abrupt and strong transition of resistivity at a predetermined temperature (related to a structural transition—melting in case of semicrystalline polymers, glass transition for resins).

In most cases, it is observed that the critical temperature T_c at which the conductor-insulator transition occurs is close to a transition temperature (melting or glass transition), at which the polymer matrix undergoes a sudden expansion, producing the interruption of the conductive paths [24]. However, for a number of CPC materials, such as those with metal fillers, PTC (resistivity spike) effects have been observed at temperatures significantly different from the transition temperatures of the respective matrices. Such examples can be CPCs with Ag-metallized glass beads filler and polymethyl methacrylate (PMMA) matrix [61], as well as CPCs with Ni-coated graphite filler and polycarbonate (PC) + polycaprolactone (PC) matrix [62], the phenomenon being explained by the large difference between the coefficients of thermal expansion of the conductive filler and the polymer matrix. According to the results of Rybak et al. [63], in the case of nanocomposites with Ag and single matrices of HDPE, or polybutylene terephthalate (PBT), as well as similar nanocomposites with binary matrices HDPE+PBT or HDPE+ Poly(m-xylene adipamide) (MXD6), the T_c value increases substantially with the content of Ag nanoparticles, being between 45 and 180°C. For example, for HDPE-xAg nanocomposites (x = volume percentage), T_c values of 44°C for x = 18% and almost 100 °C for x = 24% Ag. Similar effects were observed for PBT-xAg composites, but at higher temperatures, ranging from about 130 to 175°C, depending on the Ag concentration. Such materials would allow interesting applications as temperature sensors, since the range of maximum sensitivity can be tuned by changing some compositional parameters, such as the nature of the polymer matrix and/or the nature and concentration of the conductive filler.

Despite the fact that composite materials (especially those with thermoplastic or elastomeric matrices) are suitable for industrial applications due to their easy processing, low density, flexibility, and toughness, their use as temperature sensors has been delayed by the poor reproducibility of the resistance values of the materials subjected to repeated heating-cooling cycles in which the temperature of reaching the maximum resistivity (which corresponds to the melting temperature of the polymer matrix of thermoplastic polymers) is exceeded. Thus, the electrical resistance values can differ significantly from one heating cycle to another [50], the phenomenon being explained by the random restoration of the conductive paths during the solidification of the melt. The phenomenon can be effectively countered by radio-induced cross-linking [24, 50], as well as by using polymer mixtures as polymer matrices and/or of mixtures of conductive phases [50]. It should be noted, however, that in the case of heating elements and overcurrent protections, this problem is not so serious, since in practice the material does not reach the melting temperature, the flow of current being practically cut off before the melting temperature is reached [50].

It is also worth to mention that, in the case of thermistors, the obtaining technology is not as simple as in the case of self-regulating heating elements, since even the production of "classical" thin films of uniform thickness (~0.3 mm) is complicated by the high viscosity of composite material melts [24].

3.1.2 Thermistor performance evaluation

Evaluation of the temperature sensing behavior of PTC sensors can be done simply by placing the sensor in a programmable temperature enclosure (which allows heating from ambient temperature to a specific temperature of interest using a heating schedule, usually linear, and which also allows cooling to ambient temperature). The temperature sensor is connected to an electrical resistance measuring instrument, which allows real-time measurements in the range 0 ohm ... T ohm, or more (to be able to detect the maximum resistance at the critical transition temperature). A calibrated contact thermometer (or equivalent) measures the temperature on the sensor surface

(T_t). The dependence curve of the sensor signal (resistance and resistivity) is drawn as a function of the temperature T_t, which, in the case of the existence of a critical temperature in the scanned thermal domain, has the form of **Figure 5**.

To evaluate the performance of resistive temperature sensors, the normalized change in resistance (R_n) can be used, as well as the temperature coefficient of resistance (TCR), defined as follows [3]:

$$R_n(\%) = \frac{R - R_0}{R} \bullet 100 \tag{2}$$

$$TCR = \frac{R - R_0}{R_0} \bullet \frac{1}{\Delta T} \tag{3}$$

where R and R_0 are, respectively, the current resistance and the room temperature resistance and ΔT is the corresponding temperature interval.

It is observed that the signs of the magnitudes R_n and TCR also give the sense of resistance variation with temperature in the considered range: if R_n, TCR < 0, the resistance decreases with increasing temperature, and the material exhibits NTC effect; if R_n, TCR > 0, the material will be PTC.

3.1.3 Examples

Shafiei et al. [28] reported that the preparation of HDPE matrix composites with carbon black filler (18%) and graphene platelets (1%) showed a sudden increase in resistivity between 105 and 120°C, good repeatability, and reproducibility, showing good potential for use as a thermometer, temperature sensor, and heating elements with self-temperature regulation.

Go et al. [60] reported the obtaining of PTC composites with ethylene-vinyl acetate (EVA) matrix and CB filler (0D filler) and exfoliated Gr (2D filler) exhibiting improved intensity and reproducibility at repeated thermal cycling through mobility control filling and thermal expansion due to the combination of fillings. Additionally, these composites exhibited a temperature sensitivity approximately 14 times higher than that reported in the literature for other temperature sensors. The PTC composite with the synergistic combination of 0D and 2D fillers can detect human skin temperature by real-time monitoring and exhibited an accuracy of 0.41°C, thus demonstrating the feasibility of the PTC temperature sensor in specific applications that require sensitivity and relatively high-temperature flexibility, such as monitoring human body temperature.

Polyvinylidene fluoride (PVdF) matrix composites filled with *in situ* thermally reduced graphene oxide (TrGO) and silver nanowires (AgNW) were prepared using solution mixing followed by coagulation and hot thermal pressing [64]. Binary TrGO/PVdF nanocomposites exhibited a low percolation threshold of 0.12 vol % and a low electrical conductivity of about 10^{-7} S/cm. Blending TrGO with silver nanowires led to a significant improvement in electrical conductivity due to the synergistic effect in conductivity of the two conductive materials (the bulk conductivity of TrGO + AgNW materials was higher than the combined conductivity of TrGO/PVdF and AgNW binary composites/PVdF at the same filler content). The hybrid composites showed an increase in resistivity with temperature (PTC), the jump in resistivity being observed at the melting temperature of PVdF. The 0.04 vol % TrGO/1 vol % AgNW/PVdF hybrid material exhibited pronounced PTC behavior, making this composite an interesting candidate for current limiting devices and temperature sensors.

An ingenious sensor is that described by [10], which consists of a poly(chlorinated propylene carbonate)-based polymer foam system filled with CB and cross-linked. This system removes the typical disadvantages related to the nonlinearity and nonmonotonicity of the resistance variation with temperature, specific to PTC composites obtained by randomly dispersing the nanofiller in the polymer matrix. During the heating, the gas bubbles in the closed pores of the foam expand, causing the reduction of the wall thickness and, implicitly, the decrease of the resistivity due to the decrease of the distance between the CB particles. The process of resistance decrease is linear with temperature, reproducible, and reversible (resistivity increases with decreasing temperature).

Lyashkov et al. [65] studied composites with tungsten oxide ceramic (WO_{3}–$_{3.0}MnO_{2}$–$_{0.5}Na_{2}O_{5.0}MoO_{3}$) and polyethylene matrix with ceramic filler volume fraction from 10 to43%. That ceramic was chosen among several WO_3-based materials as having the most nonlinear I-V (*Intensity-Voltage*) characteristic. The obtained composites proved to be isotropic mixtures of filler grains in the polymer matrix and showed high values of the temperature coefficient of strength, in the range of 40–75°C, which depend on both the volume fraction of the filler and the intensity of the electric field. This dependence on the electric field can be explained by the nonlinearity of the I-V characteristic, typical for varistors. In the case of ordinary thermistors, the I-V characteristic is linear, the TCR being independent of the electric field. The dependence of the electrical conductivity of the composite on the volume fraction of the conducting ceramics can be described with a three-dimensional percolation model for a two-phase system.

3.2 Sensors containing a conductive composite with thermoset matrix

Han et al. [2] reported the obtaining of a flexible, high mechanical strength GnPs/ epoxy resin composite film exhibiting an electrical conduction percolation threshold of 1.08% (vol) GnPs. Compared to the neat polymer, the composite shows improved mechanical properties (Young's modulus +1344%, tensile strength +66.7%, and good electrical response to bending or twisting up to 180°). The percolation threshold value calculated for the film of this composite (1.08% by vol.) corresponds to the formation of a global network of connected GnPs inside the epoxy matrix, which ensures the mobility of electrons in the matrix and, implicitly, the insulator/(semi)conductor transformation of the material. After overcoming the percolation threshold, the electrical conductivity of the material increases linearly and steadily with increasing GnPs content due to the creation of numerous conductive paths through the composite. Thus, at a content of 10% (vol.) GnPs, the composite exhibits an electrical conductivity of ~0.01 S/cm, which corresponds to an increase of 11 orders of magnitude compared to the initial resin. The film acts exclusively as a temperature sensor at T > 20 °C, having a linear and stable resistive response in the range of 20–110 °C; the temperature coefficient of resistivity is 0.0063 $°C^{-1}$, higher than the standard Pt-based temperature sensor (0.0039 $°C^{-1}$[7]).

3.3 Sensors obtained through additive manufacturing technologies

Additive manufacturing (AM) technologies have completely changed the approach to R&D and production problems in various fields, such as electronics, aerospace technologies, biomedical applications, wearable technologies, and automotive industry. AM enables the faster translation of projects into marketable industrial products starting from the conceptualization of a three-dimensional (3D) model and reaching the manufacture of printed landmarks. Since AM technologies do not require tools and molds for making landmarks, as is the case for classical technologies, it follows that the application of these technologies can lead to significant savings in materials, time and labor, as well as an increase in the quality and reproducibility of production [16]. Different additive manufacturing techniques such as DIW (direct ink writing), DLP (direct light projection), FFF (fused filament fabrication) or FDM (filament deposition modeling), SLA (stereolithography), and SLS (select laser sintering) are discussed briefly by [16].

A stretchable temperature sensor based on GnPs/PDMS composite, insensitive to mechanical deformation, made by an additive manufacturing technology (3D printing) was reported in the reference [7]. The sensor exhibits a high sensitivity in temperature detection, being characterized by a temperature coefficient of resistance value of $0.0080\ ^{\circ}C^{-1}$, practically more than twice the TCR value of the standard Pt sensor.

3.4 Sensors based on sandwich structures

Sandwich structures also fall into the category of composites; in this case, the functionality of the material is given by the layers of overlapping materials. Chen et al. [3] recently reported the realization of a flexible sandwich temperature sensor from laser-reduced graphene oxide (LrGO) deposited on a PET support (**Figure 6**). The critical parameters of the sensor fabrication process are the concentration of the aqueous GO solution and the distance between the laser scan lines. For GO reduction, the minimum laser power density must exceed the GO reduction threshold but not reach the level at which ablation of the LrGO layer and PET substrate occurs. A power of 6.5 W and a scanning speed of 2000 m/s were chosen for this purpose for the UV laser with λ = 355 nm. The flexible sensor can be bent on curved surfaces, thus enabling *in situ* temperature measurement, is applicable for monitoring human breathing and space-temporal temperature variation on curved surfaces, and has great potential for realizing noncontact human-machine interfaces.

3.5 Other examples

3.5.1 Wearable sensors

Many of the applications of CNT/polymer composites are as strain sensors [2], but the association between polymer materials and CNTs can produce other interesting

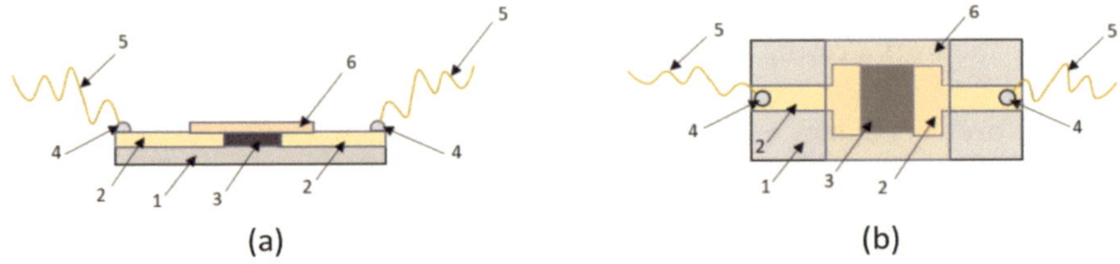

Figure 6.
*Structure of a sandwiched temperature sensor based on laser-reduced graphene oxide (adaptation after R. Chen et al. 2022): 1—Support polymer film (PET, 0.125 nm thickness); 2—T-shaped gold electrodes (Au/Ti 30/20 nm, deposited by sputtering); 3—Laser-reduced (*in situ*) graphene oxide layer (deposited from aqueous solution); 4—Conductive silver paint (for wire/electrode soldering); 5—Wire; 6 – PI tape (for sensor packaging): a—Side view; b—Top view.*

effects for emerging applications. Textiles, as assemblies of fibrous materials, single or multiple, have properties that depend on the nature of the fibers, the treatments applied to them, as well as the method of obtaining—weaving, knitting, or felting (nonwoven textiles). The integration of CNTs in textiles enables the development of wearable technologies and smart textiles, with customizable properties and functionality, through recent approaches to the synthesis of CNT-textile fiber hybrid materials. Such textile materials possess a number of important advantages over ordinary textile materials, such as low weight, integrated nonelectronic regulation of body temperature, self-cleaning without water, and appearance adapted to the requirements of clothing production [66]. Kubley et al. [66] presented a way to synthesize a sheet of carbon nanotube hybrid (CNTH) and tested ways to integrate it into fabrics for various applications, as well as their potential applications in technical and smart textiles.

Continuous monitoring of body temperature is developing rapidly, based on numerous innovations [67]. The use of additive manufacturing techniques enables the large-scale production of flexible temperature sensors, which is now a well-defined direction of development [16, 67]. Another direction is represented by bio-inspired materials and structures, such as octopus legs that have a high ability to attach to skin or other supports [19]. The sensor is resistive type and consists of a hydrogel composite formed by a poly(N-isopropylacrylamide) matrix in which PEDOT:PSS and CNTs are as conducting phases. The NTC sensor has high sensitivity between 25 and 40°C, allowing accurate detection of temperature differences of 0.5 °C and can be used in skin attachment, wearable medical, and health care applications.

3.5.2 Robotic elements with resistive sensors

In robotics, an important concern is the development and production of complex materials that reproduce the functionality of human skin (the so-called humanoid artificial skin), for the development of robots with close human characteristics, as well as for reparative surgery applications. In order to imitate the functionality of human skin, in robotics or prosthetics applications, two complex functions must be ensured, namely (i) detecting the characteristics of the surrounding environment (temperature, hardness, and slip), and the second, (ii) is to grasps various objects to move/condition them. Both of these functions require the existence of a matrix of tactile, force, and temperature sensors connected to a central acquisition system, where the image (map) of temperature and roughness of the environment with which the skin comes into contact will be obtained. Corresponding to the function (i) and based on the environmental information, the force necessary to grasp the target object will be dosed or, if the temperature is outside the established safety range (function II), it will be decided to avoid the contact with it. In addition, the sensor array (just like the skin) must have high elasticity, allowing for bending-rotating movements similar to those performed by the human hand. Obviously, practical realization involves the use of advanced manufacturing technologies such as additive manufacturing and the use of micromechanical systems (MEMS) [40].

Harada et al. [68] made a tactile force sensor (on three axes) in the form of a double 3x3 matrix, which contains a network of temperature sensors and a network of strain sensors. The device obtained exclusively through printing technologies allows the detection of sliding/frictional forces, the sense of touch (pressure), the detection of temperature, as well as the grasping/holding of objects. The temperature sensor (NTC type) consists of a printed CNT/PEDOT:PSS composite, with silver electrodes, and presents a temperature coefficient of $-0.25\%C^{-1}$, in the range of 20–80 °C.

Many of the temperature sensors used to obtain artificial skin exhibit NTC effect [10, 18, 68]. Nuthalapati et al. [69] reported a temperature sensor made of a highly sensitive rGo-Pd/kapton composite, which exhibits NTC effect at Pd contents lower than 1:4 (rGo:Pd) and PTC effect at ratios of 1:6 and 1:8. However, the sensitivity to temperature detection decreases as the Pd content increases.

4. Conclusions

Temperature sensors are key elements in the development of new innovative technologies in actual areas such as wireless health care devices, robotics, smart manufacturing, and smart products. Polymer composite materials with PTC effect are smart materials, with intrinsically abilities of self-limitation of current/voltage and, on this basis, being able to function as smart temperature self-regulating heating elements or as smart fuses. The use of PTC materials as temperature sensors (thermometers) seems complicated by the important nonlinearity of the temperature dependence of the resistivity, but over quite wide ranges, for example, between the ambient temperature and the onset of the resistance steep rise before the critical temperature, the resistance—temperature signal is practically linear enabling the development of interesting applications such as interactive monitoring body temperature with wireless devices.

Conductive carbonaceous materials such as CB, graphite, graphene, CNT, and CF are intensively studied to obtain PTC or NTC resistive temperature sensors in efficient printing-based technologies, to obtain flexible or stretchable sensors, required by new technologies of sensing. The large number of articles published in recent years in relevant publications is a further proof of the topicality of this field.

Acknowledgements

The financial support was provided by the Ministry of Research, Innovation, and Digitization through contracts: 446PED/2020, 612PED/2022, PN23140201/2023, and 25PFE/2021.

Author details

Radu Setnescu[1,2] and Eduard-Marius Lungulescu[2]*

1 Department of Sciences, Valahia University of Targoviste Faculty of Sciences and Arts, Targoviste, Romania

2 Department of Advanced Materials, INCDIE ICPE-CA (National Institute for Research and Development in Electrical Engineering), Bucharest, Romania

*Address all correspondence to: marius.lungulescu@icpe-ca.ro

References

[1] Bezzon VDN, Montanheiro TLA, de Menezes BRC, Ribas RG, Righetti VAN, Rodrigues KF, et al. Carbon nanostructure-based sensors: A brief review on recent advances. Advances in Materials Science and Engineering. 2019; **2019**:4293073

[2] Han S, Chand A, Araby S, Cai R, Chen S, Kang H, et al. Thermally and electrically conductive multifunctional sensor based on epoxy/graphene composite. Nanotechnology. 2020;**31**(7): 075702

[3] Chen R, Luo T, Geng D, Shen Z, Zhou W. Facile fabrication of a fast-response flexible temperature sensor via laser reduced graphene oxide for contactless human-machine interface. Carbon. 2022;**187**:35-46

[4] Kanao K, Nakata S, Arie T, Akita S, Takei K. Human-interactive multi-functional electronic wallpaper integrated with sensors and memory. Materials Horizons. 2017;**4**(6): 1079-1084

[5] Shih W-P, Tsao L-C, Lee C-W, Cheng M-Y, Chang C, Yang Y-J, et al. Flexible temperature sensor array based on a graphite-polydimethylsiloxane composite. Sensors. 2010;**10**(4): 3597-3610

[6] Trung TQ, Lee N-E. Flexible and stretchable physical sensor integrated platforms for wearable human-activity Monitoringand personal healthcare. Advanced Materials. 2016;**28**(22): 4338-4372

[7] Wang Z, Gao W, Zhang Q, Zheng K, Xu J, Xu W, et al. 3D-printed graphene/ polydimethylsiloxane composites for stretchable and strain-insensitive temperature sensors. ACS Applied Materials & Interfaces. 2019;**11**(1): 1344-1352

[8] Dankoco MD, Tesfay GY, Benevent E, Bendahan M. Temperature sensor realized by inkjet printing process on flexible substrate. Materials Science and Engineering: B. 2016;**205**:1-5

[9] Chen J, Yu Q, Cui X, Dong M, Zhang J, Wang C, et al. An overview of stretchable strain sensors from conductive polymer nanocomposites. Journal of Materials Chemistry C. 2019; 7(38):11710-11730

[10] Chen J, Zhu Y, Guo Z, Nasibulin AG. Recent progress on thermo-electrical properties of conductive polymer composites and their application in temperature sensors. Engineered Science. 2020;**12**:13-22

[11] Deng H, Lin L, Ji M, Zhang S, Yang M, Fu Q. Progress on the morphological control of conductive network in conductive polymer composites and the use as electroactive multifunctional materials. Progress in Polymer Science. 2014;**39**(4):627-655

[12] Ma P-C, Siddiqui NA, Marom G, Kim J-K. Dispersion and functionalization of carbon nanotubes for polymer-based nanocomposites: A review. Composites Part A: Applied Science and Manufacturing. 2010; **41**(10):1345-1367

[13] Yang R, Wang S, Zhao K, Li Y, Li C, Xia Y, et al. Comparison of oxidation polymerization methods of thiophene in aqueous medium and its mechanism. Polymer Science, Series B. 2017;**59**(1): 16-27

[14] Kumar DR, Lidster BJ, Adams RW, Turner ML. Understanding the

microstructure of poly(p-phenylenevinylene)s prepared by ring-opening metathesis polymerization using 13C-Labeled Paracyclophanediene monomers. Macromolecules. 2018; **51**(12):4572-4577

[15] Prunet G, Pawula F, Fleury G, Cloutet E, Robinson AJ, Hadziioannou G, et al. A review on conductive polymers and their hybrids for flexible and wearable thermoelectric applications. Materials Today Physics. 2021;**18**:100402

[16] Ryan KR, Down MP, Hurst NJ, Keefe EM, Banks CE. Additive manufacturing (3D printing) of electrically conductive polymers and polymer nanocomposites and their applications. eScience. 2022;**2**(4):365-381

[17] Jeon J, Lee H-B-R, Bao Z. Flexible wireless temperature sensors based on Ni microparticle-filled binary polymer composites. Advanced Materials. 2013; **25**(6):850-855

[18] Bae GY, Han JT, Lee G, Lee S, Kim SW, Park S, et al. Pressure/temperature sensing bimodal electronic skin with stimulus discriminability and linear sensitivity. Advanced Materials. 2018;**30**(43):1803388

[19] Oh JH, Hong SY, Park H, Jin SW, Jeong YR, Oh SY, et al. Fabrication of high-sensitivity skin-attachable temperature sensors with bioinspired microstructured adhesive. ACS Applied Materials & Interfaces. 2018;**10**(8): 7263-7270

[20] Katerinopoulou D, Zalar P, Sweelssen J, Kiriakidis G, Rentrop C, Groen P, et al. Large-area all-printed temperature sensing surfaces using novel composite thermistor materials. Advanced Electronic Materials. 2019; **5**(2):1800605

[21] Tan HW, An J, Chua CK, Tran T. Metallic nanoparticle inks for 3D printing of electronics. Advanced Electronic Materials. 2019;**5**(5):1800831

[22] Aguilar Ventura I, Zhou J, Lubineau G. Drastic modification of the piezoresistive behavior of polymer nanocomposites by using conductive polymer coatings. Composites Science and Technology. 2015;**117**:342-350

[23] Li Y, Huang X, Zeng L, Li R, Tian H, Fu X, et al. A review of the electrical and mechanical properties of carbon nanofiller-reinforced polymer composites. Journal of Materials Science. 2019;**54**(2):1036-1076

[24] Makuuchi K, Cheng S. Radiation Processing of Polymer Materials and its Industrial Applications. Hoboken, New Jersey, USA: John Wiley & Sons; 2012

[25] Huang J-C. Carbon black filled conducting polymers and polymer blends. Advances in Polymer Technology. 2002;**21**(4):299-313

[26] Mora A, Han F, Lubineau G. Estimating and understanding the efficiency of nanoparticles in enhancing the conductivity of carbon nanotube/polymer composites. Results in Physics. 2018;**10**:81-90

[27] Liu Y, Zhang H, Porwal H, Busfield JJ, Peijs T, Bilotti E. Pyroresistivity in conductive polymer composites: A perspective on recent advances and new applications. Polymer International. 2019;**68**(3): 299-305

[28] Shafiei M, Ghasemi I, Gomari S, Abedini A, Jamjah R. Positive temperature coefficient and electrical conductivity investigation of hybrid nanocomposites based on high-density polyethylene/graphene Nanoplatelets/

carbon black. Physica status solidi (a). 2021;**218**(20):2100361

[29] Pang H, Xu L, Yan D-X, Li Z-M. Conductive polymer composites with segregated structures. Progress in Polymer Science. 2014;**39**(11):1908-1933

[30] Xie L, Zhu Y. Tune the phase morphology to design conductive polymer composites: A review. Polymer Composites. 2018;**39**(9):2985-2996

[31] Strugova D, Ferreira Junior JC, David É, Demarquette NR. Ultra-low percolation threshold induced by thermal treatments in Co-continuous blend-based PP/PS/MWCNTs nanocomposites. Nanomaterials. 2021; **11**(6):1620

[32] Tanaka K. Chapter 1 - classification of carbon. In: Tanaka K, Iijima S, editors. Carbon Nanotubes and Graphene. Second ed. Oxford: Elsevier; 2014. pp. 1-5

[33] Gao Y, Tykwinski RR. Advances in Polyynes to model Carbyne. Accounts of Chemical Research. 2022;**55**(24): 3616-3630

[34] Eisler S, Slepkov AD, Elliott E, Luu T, McDonald R, Hegmann FA, et al. Polyynes as a model for Carbyne: Synthesis, physical properties, and nonlinear optical response. Journal of the American Chemical Society. 2005; **127**(8):2666-2676

[35] Artyukhov VI, Liu M, Yakobson BI. Mechanically induced metal–insulator transition in carbyne. Nano Letters. 2014;**14**(8):4224-4229

[36] Khoo KH, Neaton JB, Son YW, Cohen ML, Louie SG. Negative differential resistance in carbon atomic wire-carbon nanotube junctions. Nano Letters. 2008;**8**(9):2900-2905

[37] Kaiser AB, Skákalová V. Electronic conduction in polymers, carbon nanotubes and graphene. Chemical Society Reviews. 2011;**40**(7):3786-3801

[38] Klauk H. Organic thin-film transistors. Chemical Society Reviews. 2010;**39**(7):2643-2666

[39] Dekker C. Carbon nanotubes as molecular quantum wires. Physics Today. 1999;**52**(5):22-28

[40] Kolahdouz M, Xu B, Nasiri AF, Fathollahzadeh M, Manian M, Aghababa H, et al. Carbon-related materials: Graphene and carbon nanotubes in semiconductor applications and design. Micromachines. 2022;**13**(8):1257

[41] Niyobuhungiro D, Hong L. Graphene polymer composites: Review on fabrication method, properties and future perspectives. Advances in Science and Technology Research Journal. 2021; **15**(1):37-49

[42] Nwosu CN, Iliut M, Vijayaraghavan A. Graphene and water-based elastomer nanocomposites – A review. Nanoscale. 2021;**13**(21): 9505-9540

[43] Lee SJ, Yoon SJ, Jeon I-Y. Graphene/polymer nanocomposites: Preparation, mechanical properties, and application. Polymers. 2022;**14**(21):4733

[44] Priyadarsini S, Mohanty S, Mukherjee S, Basu S, Mishra M. Graphene and graphene oxide as nanomaterials for medicine and biology application. Journal of Nanostructure in Chemistry. 2018;**8**(2):123-137

[45] Araby S, Meng Q, Zhang L, Kang H, Majewski P, Tang Y, et al. Electrically and thermally conductive elastomer/graphene nanocomposites by solution mixing. Polymer. 2014;**55**(1):201-210

[46] Meng Q, Jin J, Wang R, Kuan H-C, Ma J, Kawashima N, et al. Processable 3-nm thick graphene platelets of high electrical conductivity and their epoxy composites. Nanotechnology. 2014;**25**(12):125707

[47] He XJ, Du JH, Ying Z, Cheng HM, He XJ. Positive temperature coefficient effect in multiwalled carbon nanotube/high-density polyethylene composites. Applied Physics Letters. 2005;**86**(6):062112

[48] Zeng Y, Lu G, Wang H, Du J, Ying Z, Liu C. Positive temperature coefficient thermistors based on carbon nanotube/polymer composites. Scientific Reports. 2014;**4**(1):6684

[49] Zhou J, Lubineau G. Improving electrical conductivity in polycarbonate nanocomposites using highly conductive PEDOT/PSS coated MWCNTs. ACS Applied Materials & Interfaces. 2013; **5**(13):6189-6200

[50] Setnescu R, Lungulescu E-M, Marinescu VE. Polymer composites with self-regulating temperature behavior: Properties and characterization. Materials. 2022;**16**(1):157

[51] Amjadi M, Kyung K-U, Park I, Sitti M. Stretchable, skin-mountable, and wearable strain sensors and their potential applications: A review. Advanced Functional Materials. 2016; **26**(11):1678-1698

[52] Zha J-W, Wu D-H, Yang Y, Wu Y-H, Li RKY, Dang Z-M. Enhanced positive temperature coefficient behavior of the high-density polyethylene composites with multi-dimensional carbon fillers and their use for temperature-sensing resistors. RSC Advances. 2017;**7**(19):11338-11344

[53] Lee J-H, Kim SK, Kim NH. Effects of the addition of multi-walled carbon nanotubes on the positive temperature coefficient characteristics of carbon-black-filled high-density polyethylene nanocomposites. Scripta Materialia. 2006;**55**(12):1119-1122

[54] Yang Y, Yuan C-l, Chen G-h, Yang T, Luo Y, Zhou C-r. Effect of Ba0.5Bi0.5Fe0.9Sn0.1O3 addition on electrical properties of thick-film thermistors. Transactions of Nonferrous Metals Society of China. 2015;**25**(12): 4008-4017

[55] Choi H-J, Kim MS, Ahn D, Yeo SY, Lee S. Electrical percolation threshold of carbon black in a polymer matrix and its application to antistatic fibre. Scientific Reports. 2019;**9**(1):6338

[56] Chen J, Cui X, Sui K, Zhu Y, Jiang W. Balance the electrical properties and mechanical properties of carbon black filled immiscible polymer blends with a double percolation structure. Composites Science and Technology. 2017;**140**:99-105

[57] Sánchez-Sánchez X, Elias-Zuñiga A, Hernández-Avila M. Processing of ultra-high molecular weight polyethylene/ graphite composites by ultrasonic injection moulding: Taguchi optimization. Ultrasonics Sonochemistry. 2018;**44**:350-358

[58] Forintos N, Czigany T. Multifunctional application of carbon fiber reinforced polymer composites: Electrical properties of the reinforcing carbon fibers – A short review. Composites Part B: Engineering. 2019; **162**:331-343

[59] Sankaran S, Deshmukh K, Ahamed MB, Khadheer Pasha SK. Recent advances in electromagnetic interference shielding properties of metal and carbon filler reinforced flexible polymer composites: A review. Composites Part

A: Applied Science and Manufacturing. 2018;**114**:49-71

[60] Go G-M, Park S, Lim M, Jang B, Park JY, Cho H-B, et al. Enhanced positive temperature coefficient intensity and reproducibility with synergistic effect of 0-D and 2-D filler composites. Journal of Materials Science. 2022;**57**(38):18037-18050

[61] Kar P, Khatua BB. Highly reversible and repeatable PTCR characteristics of PMMA/Ag-coated glass bead composites based on CTE mismatch phenomena. Polymer Engineering & Science. 2011; **51**(9):1780-1790

[62] Kar P, Khatua BB. PTCR characteristics of polycarbonate/nickel-coated graphite-based conducting polymeric composites in presence of poly(caprolactone). Polymer Composites. 2011;**32**(5):747-755

[63] Rybak A, Boiteux G, Melis F, Seytre G. Conductive polymer composites based on metallic nanofiller as smart materials for current limiting devices. Composites Science and Technology. 2010;**70**(2):410-416

[64] He L, Tjong S-C. Electrical behavior and positive temperature coefficient effect of graphene/polyvinylidene fluoride composites containing silver nanowires. Nanoscale Research Letters. 2014;**9**(1):375

[65] Lyashkov AY, Makarov VO, Plakhtii YG. Structure and electrical properties of polymer composites based on tungsten oxide varistor ceramics. Ceramics International. 2022;**48**(6): 8306-8313

[66] Kubley A, Chitranshi M, Hou X, Schulz M. Manufacturing and characterization of customizable flexible carbon nanotube fabrics for smart wearable applications. Text. 2021;**1**(3): 534-546

[67] Su Y, Ma C, Chen J, Wu H, Luo W, Peng Y, et al. Printable, highly sensitive flexible temperature sensors for human body temperature monitoring: A review. Nanoscale Research Letters. 2020;**15**(1): 200

[68] Harada S, Kanao K, Yamamoto Y, Arie T, Akita S, Takei K. Fully printed flexible fingerprint-like three-Axis tactile and slip force and temperature sensors for artificial skin. ACS Nano. 2014;**8**(12):12851-12857

[69] Nuthalapati S, Shirhatti V, Kedambaimoole V, Pandi NV, Takao H, Nayak MM, et al. Highly sensitive flexible strain and temperature sensors using solution processed graphene palladium nanocomposite. Sensors and Actuators A: Physical. 2022;**334**:113314

Chapter 7

Current Status and State-of-Art Developments in Temperature Sensor Technology

Deqi Chen, Qianlong Zuo, Hao Wu, Haidong Liu and Fenglei Niu

Abstract

Temperature is one of the seven base units of the physical world, and the temperature sensors have wide applications in the lives, research, and industries. This chapter presents a brief introduction on four classic types of temperature sensors, including thermometers, thermocouples, resistance temperature detectors (RTD), and thermistors. These traditional temperature sensors have some limitations and are not suitable for dynamic measurements. To meet the demand for temperature measurement under various extreme and complex conditions, four advanced types of temperature sensors are introduced. The optical temperature sensors, including the infrared thermal imaging and laser temperature sensor, utilize the thermal radiation and are capable of measuring high-temperature objects without direct contact. The small and flexible fiber optic temperature sensors take advantage of the fact that the temperature plays a significant role in the optical transmission characteristics of the optical fiber, and it can be used in point, quasi-distributed, or distributed form. Acoustic temperature sensors measure the speed and frequency of the sound wave under different temperatures to obtain the temperature, and it is commonly used for health monitoring of complex structures. Furthermore, micro/nano temperature sensors are ideal for specific applications due to their small size, high sensitivity, and rapid response time.

Keywords: temperature sensor, thermocouple, optical, acoustic, micro/nano-scale

1. Introduction

Temperature is a widely recognized and essential physical quantity for measuring the hot and cold states of various systems. Its accurate measurement is critical for assessing the health and proper functioning of equipment and for deepening our understanding of physical phenomena. Thus, temperature sensing is a highly significant area of research. Generally, temperature sensors are typically fabricated based on various thermal effects such as thermoelectric, thermal expansion, thermo-optical, and thermomagnetic effects, which provide highly accurate and stable measurements in different application scenarios. By leveraging these diverse thermal effects, researchers can develop sensors with increasingly advanced capabilities to enable a wide range of scientific and technological breakthroughs.

2. Traditional temperature sensors

2.1 Expansion type temperature sensor

The thermal expansion and contraction of most substances can be harnessed for temperature measurement purposes. Depending on the physical state of the working material, temperature sensors can be categorized into gas, glass tube liquid, and solid thermometers (such as bimetallic thermometers). One such thermometer, the mercury thermometer, is a traditional and widely used glass liquid thermometer as shown in **Figure 1** based on the thermal expansion effect. By measuring the volume change of the thermometer with changes in temperature, the mercury thermometer is able to accurately determine temperature. Due to its simple structure and lack of external energy requirements, the mercury thermometer is often used as a household thermometer. Its accuracy ranges between 0.005 and 4°C, which is primarily determined by the precision of the capillary tube manufacturing process.

2.2 Thermocouple

Thermocouples represent a popular and highly effective temperature measurement technology that relies on the variation of thermoelectric potential difference generated at the contact point with temperature change. The selection of thermocouple materials is based on criteria, such as chemical stability and high thermal response electromotive force, which can impact measurement accuracy. Moreover, the accuracy of a thermocouple depends on its class of manufacturing tolerance. Common thermocouple

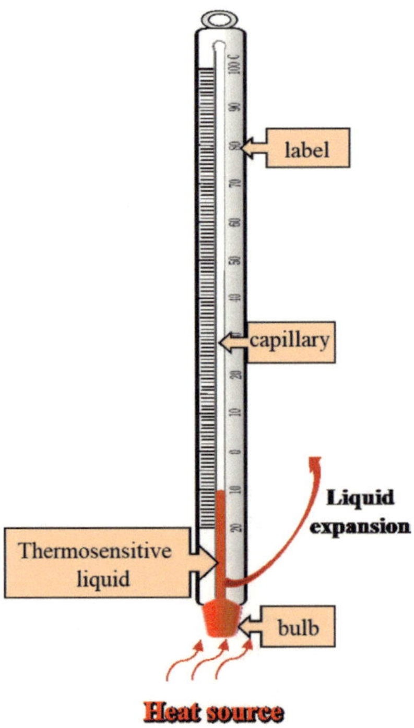

Figure 1.
The mercury thermometer.

standardizations include Type S, Type T, and Type K, which are known for their reliability and versatility in different measurement scenarios. Specific thermocouple types are shown in **Table 1**. It also demonstrates the effective measurement range and accuracy for the rank II thermocouples. With advances in measurement accuracy and remote signal transmission, thermocouples have become widely used for temperature measurement from -40°C to 1700°C. Given their widespread use and versatility, researchers have also explored various improvements to basic thermocouples, such as multi-probe type thermocouples and thin film thermocouples (TFTCs), as shown in **Figure 2**, to expand their functionality and precision in a wide range of applications.

2.3 Resistance temperature detector

A resistance temperature detector (RTD) operates on the principle that resistance varies with temperature. This type of temperature sensor provides an analog signal of resistance value when an external constant excitation current is applied. It uses the RTD's excellent linear thermal response to characterize the measured temperature value. Platinum (Pt) is the preferred material for RTD temperature sensors due to its

Type	Thermoelectric electrode material		Effective range/°C	Rank II accuracy
	Positive pole	Negative pole		
S	Pt-Rh 10	Pt	0–1600	1.5°C or 0.25%T_{max}
R	Pt-Rh 13	Pt	0–1600	1.5°C or 0.25%T_{max}
B	Pt-Rh 30	Pt-Rh 6	6–1700	1.5 °C or 0.25%T_{max}
K	Ni-Cr	Ni- Al	−40–1200	2.5 °C or 0.75%T_{max}
T	Cu	Cu-Ni	−40–350	1 °C or 0.75%T_{max}
J	Fe	Cu-Ni	−40–750	2.5 °C or 0.75%T_{max}
E	Ni-Cr	Cu-Ni	−40–900	2.5 °C or 0.75%T_{max}

Table 1.
Thermocouple types [1].

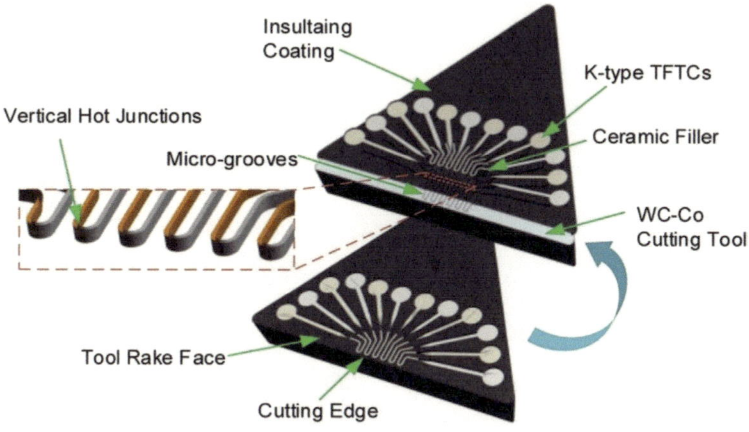

Figure 2.
Thin film thermocouple [2].

Rank	Effective temperature range/°C		Accuracy/°C
	Wire element	Film element	
AA	−50 ~ 250	0 ~ 150	±(0.1 + 0.17%|T|)
A	−100 ~ 450	−30 ~ 300	±(0.15 + 0.2%|T|)
B	−196 ~ 600	−50 ~ 500	±(0.3 + 0.5%|T|)
C	−196 ~ 600	−50 ~ 600	±(0.6 + 1%|T|)

Table 2.
Platinum resistance temperature detector [3].

excellent analog and digital linearity. Pt series RTDs have a temperature measurement range from -196°C to 600°C, and their measurement accuracy is much higher than that of thermocouples. For different RTDs within different wire-wound element and thin film element types, specific information is shown in **Table 2**. However, due to the small change in resistance with temperature, a high-precision reference and high-resolution analog-to-digital conversion (ADC) circuit are required. Therefore, platinum resistance is divided into three circuit wiring methods: two-wire, three-wire, and four-wire measurement methods. The four-wire circuit form can completely eliminate the error caused by lead resistance, as shown in **Figure 3**. It is mainly used for high-precision temperature measurement.

2.4 Thermistor

A thermistor is an inexpensive temperature measurement element that exhibits a change in resistance magnitude with temperature. Unlike an RTD, its degree of linear thermal response is low, but it has high-temperature sensitivity and is made from materials that include metals, alloys, and semiconductors. Depending on the type of resistance change with temperature, thermistors are classified as either NTC (negative temperature coefficient, exhibiting a nonlinear relationship) or PTC (positive temperature coefficient). Due to their simple structure, thermistors are also widely used in the electronics industry, such as **Figure 4**.

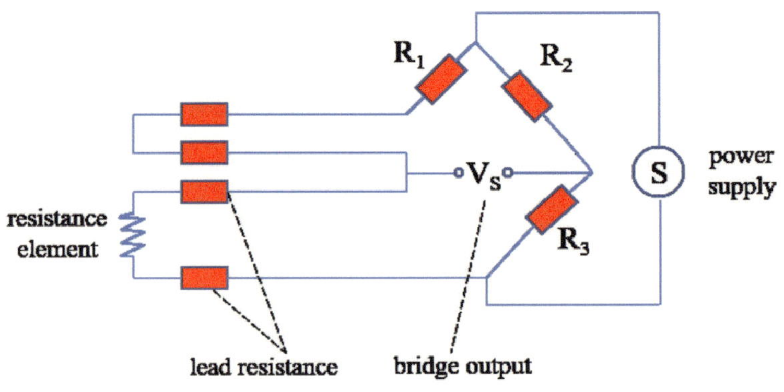

Figure 3.
4 leg measurement for RTD.

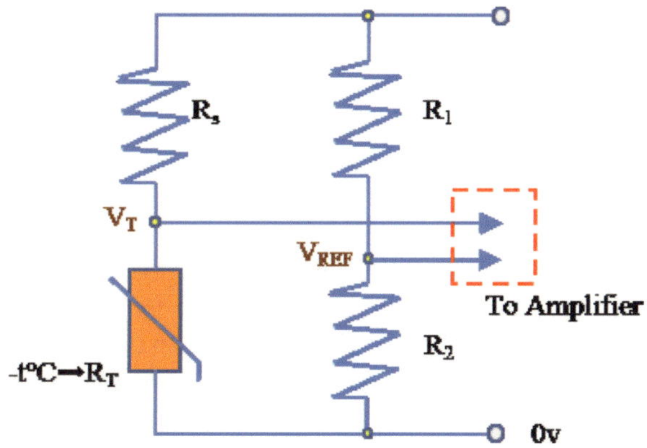

Figure 4.
Thermistor schematic.

2.5 Summary

Traditional temperature sensors have been widely used in daily life and industry, but they have gradually revealed certain limitations during their use. For example, thermal expansion-based sensors, such as the mercury thermometer, are prone to high-temperature failures and have poor universality in industrial applications. Additionally, these sensors are not suitable for dynamic measurements. Although the thermocouple, RTD, and thermistor measurement methods have overcome some of these limitations, they are still contact monitoring methods and have certain short-comings such as point contact measurement, complex wiring systems, measurement errors caused by equipment self-heating during the process, and unsuitability for high-temperature extreme operating conditions.

3. Advanced temperature sensors

In addition to the well-known disadvantages mentioned above, conventional temperature sensors face limitations when it comes to measuring temperature in complex and changing harsh environments or obtaining microscopic temperature data. In order to overcome these limitations and meet the demand for temperature measurement in extreme and complex conditions, advanced temperature sensors have emerged. These include optical temperature sensors, fiber optic temperature sensors, acoustic temperature sensors, and micro and nano temperature sensors. In this section, we will provide a detailed introduction to these advanced temperature sensors based on their typical classification characteristics.

3.1 Optical temperature sensor

Optical temperature sensors utilize the principles of Stephen Boltzmann's law and Planck's radiation law to measure temperature based on the optical effects of heat. This enables noncontact temperature measurement, allowing for the measurement of the surface temperature of high-temperature objects. Due to these advantages, optical temperature sensors are widely used in industrial production and food processing.

3.1.1 Infrared thermal imaging

An infrared thermal imager is the most widely used device among optical temperature sensors. It is based on the principle of thermal radiation of infrared to construct temperature fields. The current state-of-the-art direction in infrared thermal imager technology is reflected in the micro-electro-mechanical systems (MEMS) manufacturing process. The infrared thermal imager receives infrared radiation energy and converts it into temperature gradients through the front-end detector. The temperature gradients are then converted into electrical signals by the thermopile, and digital signals are obtained after amplification, shaping, and analog-to-digital conversion. These signals are visualized as temperature clouds on the display. Infrared thermal imaging provides an effective and fast method for real-time surface temperature measurement and is suitable for surface temperature field measurement with uneven temperature distribution or surface monitoring of superheated temperature.

The performance of IR MEMS is closely related to factors such as IR absorber absorption efficiency, thermopile performance, and thermal isolation layer material [4]. As a result, significant research efforts have been made to further enhance IR MEMS performance. The schematic of IR MEMS is shown as **Figure 5**. Key performance indicators for IR MEMS include time response, responsiveness, and noise level. To improve IR absorber absorption efficiency, Hou et al. [6] used carbon nanoparticles (CNP), Si_3N_4, and TiN nanoparticles loaded with a porous structured carbon microparticle (CMP) coating (CMP/CNP-Si_3N_4-TiN) as absorption media. The experimental results demonstrated that the coating has the capability to function as highly sensitive broadband absorbers in the range of 3–5 μm to 8–14 μm, achieving absorption rates of 93.8% and 92.6%, respectively. To improve the time response rate, Li et al. [7] utilized a novel single-sided micromachining technique fabricated in 111 wafers. They found that the p-Si-Al thermocouple in series in the IR thermopile demonstrated significantly higher Seebeck coefficients and lower noise compared to the conventional polycrystalline Si-Al thermocouple. By optimizing the cross-sectional area of the two thermoelectric material layers, the signal-to-noise ratio (D*) of the thermopile was improved, resulting in an ultrahigh responsivity of 342 V/W and an ultrashort response time of 0.56 ms.

In various fields such as fire detection, security monitoring, human body temperature measurement, drone control, and industrial automation, infrared thermal imaging technology has gained widespread application due to its fast response, low

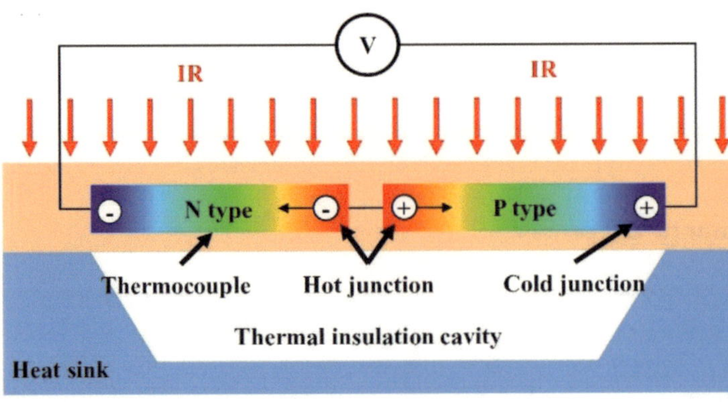

Figure 5.
Infrared thermopile components schematic [5].

noise, absence of mechanical movement, wide spectral range, high stability, and strong anti-interference capabilities.

3.1.2 Laser temperature sensor

Unlike infrared thermography, a laser temperature sensor is an active temperature measurement technology that calculates the surface temperature of an object by actively emitting a laser beam onto the surface of the object and measuring the energy reflected or scattered by the laser. Since the laser is actively emitted and received for temperature measurement, it can achieve long-distance temperature measurement and avoid the influence of environmental radiation.

In many cases, the emissivity of the object being measured is unknown, prompting the UK's National Physical Laboratory (NPL) [8] to study and propose the laser emissivity free thermometry (LEFT) technique such as **Figure 6** shown. This laser-based method for measuring target surface temperature does not require knowledge of the object's emissivity. The LEFT method is achieved by analyzing the ratio of intensities of two different laser wavelengths absorbed by the target material being tested. While the laser temperature measurement method performs well on surfaces with high emissivity, it may encounter difficulties when applied to surfaces with low emissivity. An et al. [9] constructed a noncontact temperature measurement system based on the principle of infrared laser radiation temperature measurement shown in **Figure 7**. The experimental results revealed that, for surface temperatures between

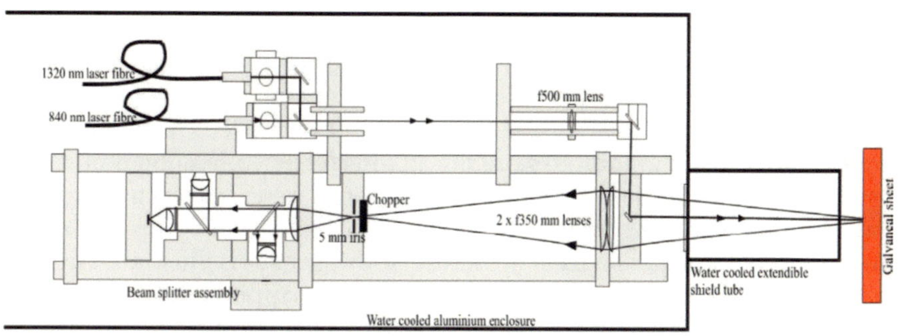

Figure 6.
LEFT device overall structure schematic [8].

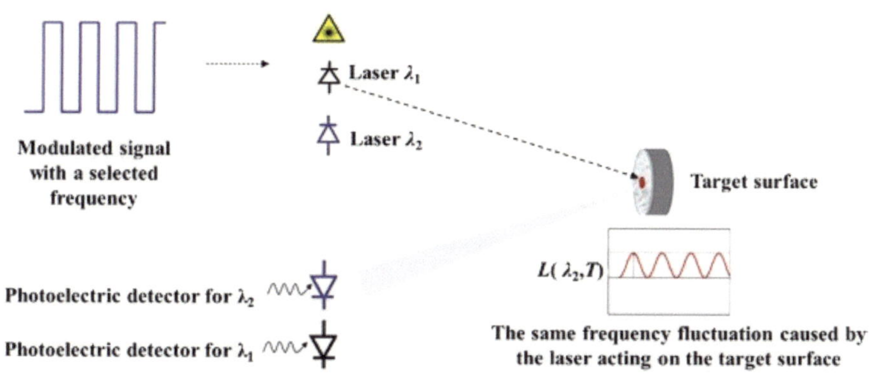

Figure 7.
Active dual wavelength infrared laser measurement schematic [9].

873 K and 1173 K, the relative deviation of the results from the reference value was within 0.5% for high emissivity samples. For low emissivity samples, the relative deviation from the reference temperature was within 0.8%, with an average absolute deviation of 3.3 K. The researchers also measured the surface temperature of low emissivity samples from 873K to 1173K.

3.2 Optical fiber temperature sensor

The fiber optic temperature sensor is a type of sensor that utilizes optical fiber transmission to measure temperature. Unlike optical temperature sensors, which operate on a different principle, fiber optic temperature sensors rely on the thermal effect and spectral properties of optical fiber to perform temperature measurements. When subjected to varying temperatures, the refractive index within the fiber structure changes, leading to alterations in the optical transmission characteristics. Due to the small and flexible structure of optical fibers, this type of temperature sensor is particularly well-suited for use in complex environments with high temperatures, pressures, and radiation levels. Additionally, the number of temperature measurement points can vary, with fiber optic temperature sensors typically categorized as point, quasi-distributed, or distributed.

3.2.1 Point temperature sensor

The point-type fiber optic temperature sensor is capable of measuring the temperature at a single point in space using its temperature measurement method. Virginia Tech [10] first explored the sapphire fiber Fabry-Perot (F-P) sensor as a point-type fiber optic temperature sensor due to its high heat resistance and measurement resolution. This sensor utilizes the thermal expansion or thermogenic effect at the temperature measurement endpoint to generate an interference phenomenon between two light beams *via* refraction or reflection. This interference causes a change in the phase and amplitude of the optical signal, which is used to evaluate the temperature at the measurement point. The basic structure of this sensor is illustrated in **Figure 8**.

Jiang's group [11] conducted a series of studies on sapphire fiber optic sensors, which resulted in the development of an all-sapphire fiber optic temperature sensor [12]. This sensor is packaged in an all-sapphire structure, as shown in **Figure 9**, which effectively eliminates the mismatch in the coefficient of thermal expansion of the material under high-temperature conditions. The device is capable of operating at high temperatures ranging from 1000 to 1500°C for extended periods while maintaining a linear sensitivity growth trend with temperature. The evaluation conducted at room temperature indicates a measurement accuracy of 0.15 μm with an error of less than 2% of the full scale. Furthermore, the team proposed a higher-order mode suppression technique based on the sapphire fiber sensor

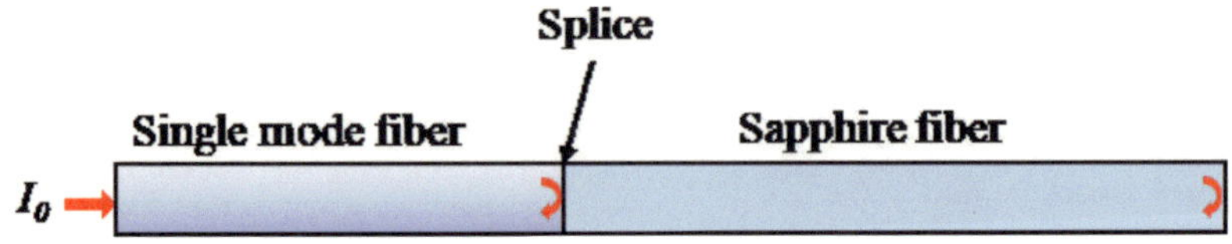

Figure 8.
Conventional sapphire fiber optic Fabry-Perot sensors [10].

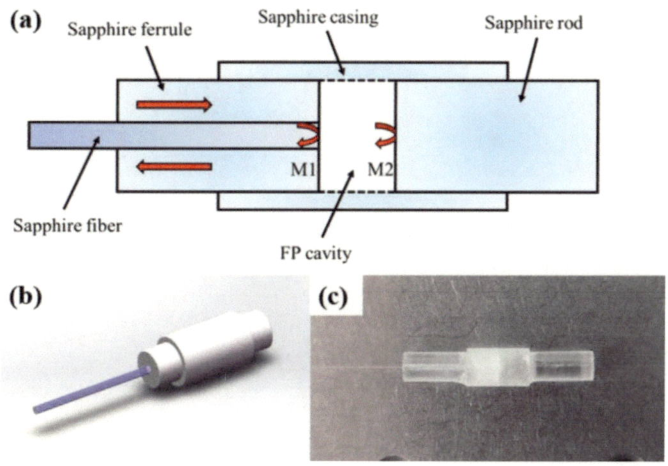

Figure 9.
All-sapphire fiber optic temperature sensor [12].

using multimode [13] or tapered multimode [14], which effectively mitigates the irregular coupling between higher-order modes and fundamental modes, thereby improving the signal-to-noise ratio of interference fringes. Due to their high accuracy, sensitivity, and long measurement distance, point-type fiber optic temperature sensors find wide application in industrial, medical, and aviation fields for temperature measurement in extreme environments characterized by high temperature, high pressure, and high radiation.

3.2.2 Quasi-distributed temperature sensor

The quasi-distributed temperature sensor employs a periodic grating fabricated on an optical fiber structure to enable temperature measurement over a distance where the grating is arranged. As the optical signal enters the optical fiber grating structure and passes through a location with a large change in refractive index between the gratings, reflection and transmission phenomena occur. The wavelength of the reflected light is influenced by the grating period and the amount of refractive index change. Therefore, by measuring the optical properties, specifically the wavelength, of the reflected light signal, the temperature value at the location of the fiber grating can be inferred, as depicted in **Figure 10**.

Thermally regenerative Bragg fiber grating sensors (RFBG) are widely used in fiber grating temperature sensors due to their high sensitivity and stability, especially their exceptional performance in high-temperature environments. The Hong Kong Polytechnic University [16] achieved the first secondary thermally regenerated Bragg fiber grating sensor (R^2FBG) by continuously ramping the temperature, enabling temperature measurements of up to 1400°C. The device exhibits not only a temperature sensitivity of 13.7 pm/°C and excellent linearity at 250–900°C but also 15.3 pm/°C at high temperatures of 900–1370°C, and even more outstanding linearity (R2 = 99.9%). In addition to its excellent high-temperature measurement performance, RFBG is also suitable for long-term temperature measurements. F.J. Dutz [17] employed four six-element RFBG arrays in a chemical test stack with the package structure shown in **Figure 11**. The device operated for two years from 150°C to 500°C and exhibited no failures or significant wavelength drift.

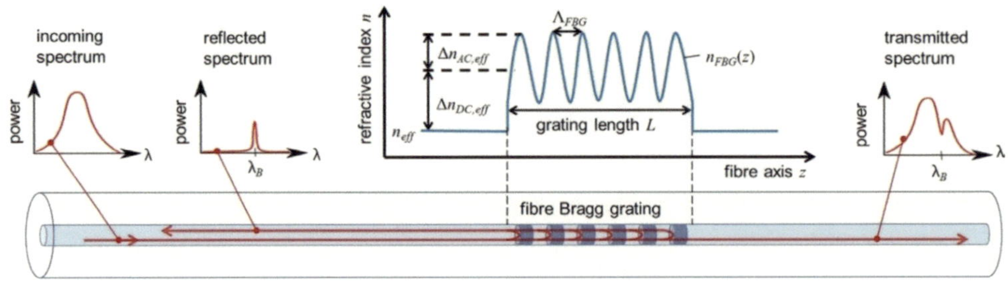

Figure 10.
Fiber Bragg grating principal schematic [15].

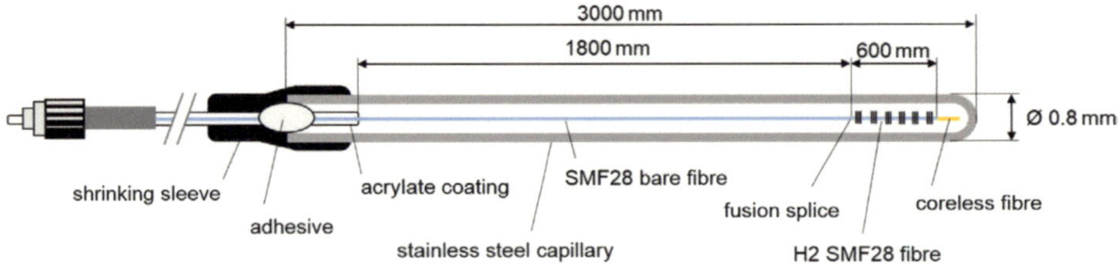

Figure 11.
Six-element regenerative Bragg grating package diagram [17].

Thermal RFBG has several advantages such as high-temperature resolution, accuracy, real-time response, and strong resistance to electromagnetic interference. Therefore, it finds extensive applications in industrial automation, forest fire monitoring, and other fields.

3.2.3 Distributed temperature sensor

Compared to quasi-distributed temperature sensors, distributed fiber optic temperature sensors enable temperature measurement over the entire length of the fiber optic path rather than just at local points. These sensors transmit optical signals through temperature-sensitive materials, such as optical fibers, and obtain photon energy and phase changes at various locations along the fiber line using Raman scattering or Brillouin scattering, ultimately allowing for the temperature distribution to be measured along the fiber. Among these sensors, the distributed temperature sensor-Raman (DTS-R) system is typically based on the optical time-domain reflectometer (OTDR) principle as shown in **Figure 12**, which involves emitting short pulses in the fiber and using the optical time difference between the round-trip to provide temperature variation and spatial location information along the entire fiber, as shown schematically below.

Marianne Stely Peixoto e Silva [19] conducted a comprehensive evaluation of the wide temperature performance of commercial time-domain optical reflectometry (OTDR) and erbium-doped fiber amplifiers (EDFA) on optical fibers up to 6 km in length. The results showed that the sensor sensitivity was 0.01 dB/°C@100 ns, providing temperature measurement from −196 to 400°C with a system resolution of 5°C. The sensor accuracy was 5°C in the range from −196 to 187°C, while it could reach 11.5°C at higher temperatures. Furthermore, Liu's group [20] explored the performance of temperature distributors based on Raman scattering at high

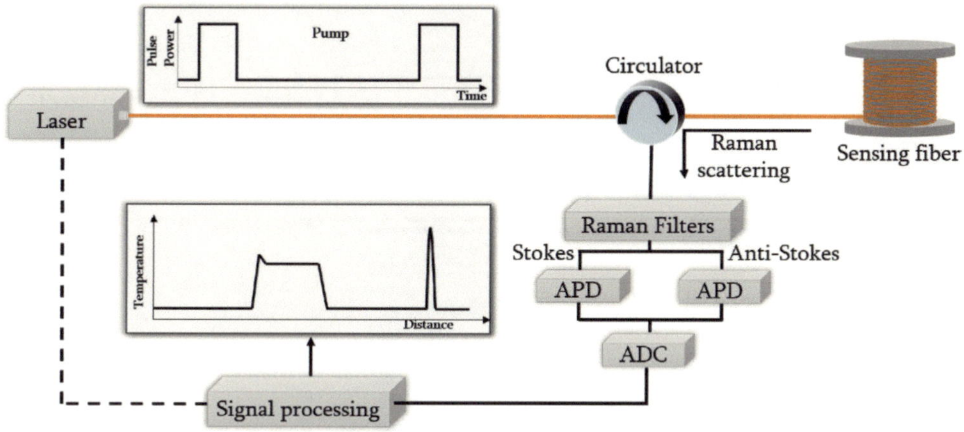

Figure 12.
Schematic diagram of Raman distributed temperature sensor system based on OTDR technique [18].

temperatures. They proposed a Raman scattering temperature distributor system based on sapphire fibers under experimental conditions, using a high-power picosecond pulsed laser with a wavelength of 532 nm, which exhibited stability at a temperature resolution of up to 1200°C, with a spatial resolution of 14 cm and a temperature resolution of 3.7°C. In a subsequent experiment [21] with a fiber length of 2 m, they used a sub-nanosecond pulsed laser to increase the temperature detection limit to 1400°C and spatial resolution to 12.4 cm. Distributed optical fiber temperature sensors are commonly used in places such as oil and gas pipelines or transmission lines that require long-distance temperature measurements due to their characteristics.

3.3 Acoustic temperature sensor

An acoustic temperature sensor detects temperature by measuring changes in the speed and frequency of sound waves. Based on the principles of acoustics, the sensor uses the relationship between the speed, propagation characteristics, and temperature of sound waves propagating in a medium to measure temperature. As the temperature in the medium changes, the speed and frequency of sound waves also change in accordance with the thermal properties of the medium. The sensor is capable of measuring these changes and calculating the temperature. Due to its unique monitoring characteristics, acoustic temperature sensors are commonly used for the health monitoring of complex structures.

3.3.1 Acoustic time-of-flight

The velocity of sound propagation is closely linked to the density, elastic modulus, and pressure of the medium it travels through. Temperature changes can be detected by observing the effect of temperature on the time-of-flight (TOF) in the medium. Typically, the speed of sound is directly proportional to temperature, such that as temperature increases, so too does the speed of sound in the medium. By measuring changes in sound velocity, temperature changes can be inferred. Wang [22] completed ultrasonic TOF measurements to determine temperature using a waveguide device. The measurement process requires a small region of effective

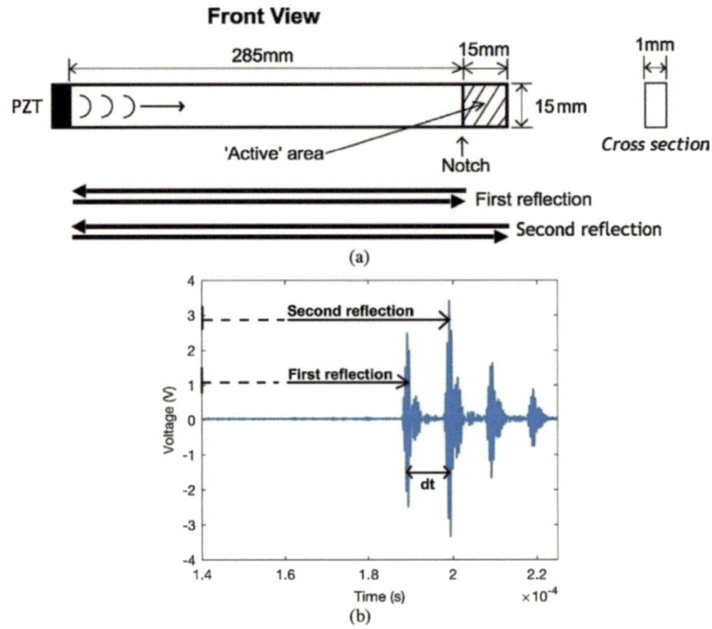

Figure 13.
Waveguide temperature measurement schematic [22].

thermal inertia to be placed within the temperature field, as illustrated in **Figure 13**. This device can achieve high-precision temperature measurement of 0.015°C in constant temperature water bath environments below 100°C (using PT100 RTD as a reference). As such, this approach is a cost-effective and reliable alternative to contact measurement methods.

In addition to contact-based temperature measurement, this principle also allows for contactless temperature restoration at the measurement point [23], as depicted in **Figure 14**. Wang [24] proposed a temperature field reconstruction algorithm that enables two-dimensional temperature field reconstruction, reducing the relative error to a maximum of 2.881%, and improving the anti-noise interference capability. This contactless acoustic measurement method provides real-time measurement, high accuracy, a wide measurement range, and environmental adaptability, making it a promising solution for temperature measurement in harsh environments.

3.3.2 Resonant frequency

This type of temperature sensor combines the principle of piezoelectric effect. The piezoelectric element is placed at the measurement position, and by applying an electric field to the piezoelectric element, the inverse piezoelectric effect of the piezoelectric element causes it to vibrate and produce sound waves, and the piezoelectric effect of the piezoelectric element generates an RF signal. When the temperature of an object is at a certain temperature, the gap width changes due to the thermal expansion effect of the piezoelectric element, and the resonance frequency changes. The detector analyzes the RF signal (the RF signal of the upper and lower electrodes is strongest at the resonance frequency) to obtain the temperature.

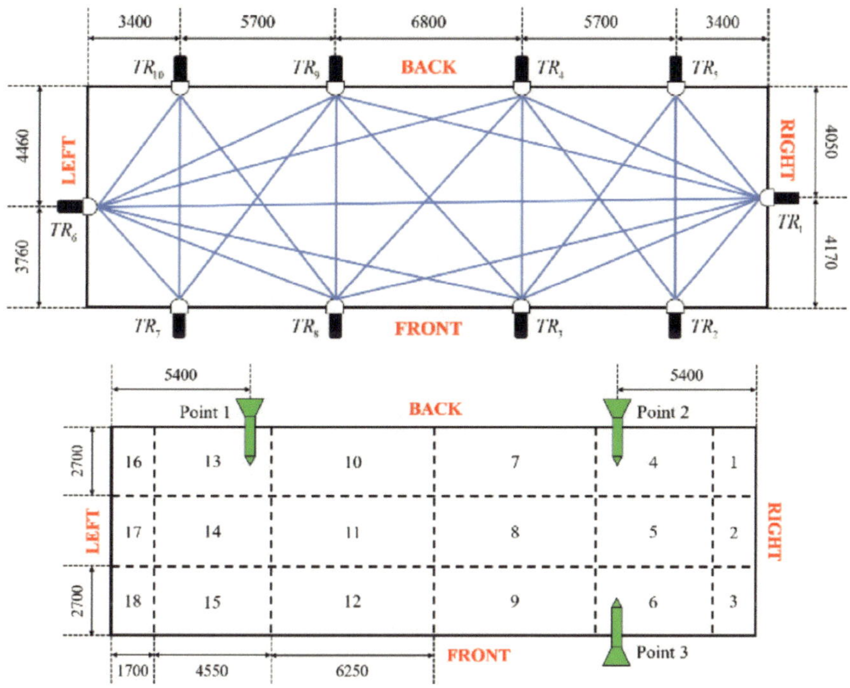

Figure 14.
TOF-based temperature field reconstruction method [23].

Despite the promising theoretical basis for thin film bulk acoustic resonator (FBAR) temperature sensors, the technology is still in the experimental study stage and has not been commercially promoted. Zhang [25] proposed an FBAR temperature sensor shown in **Figure 15** with a patterned support layer that demonstrated a differential pressure linearity error of only 0.35%, significantly smaller than existing sensors. However, the sensitivity parameter of the FBAR temperature sensor is critical, as demonstrated by the slow temperature stability achieved within 110 s and an accuracy of 0.015°C when the resonator temperature was controlled at 75°C and the ambient temperature was 25°C. To improve the temperature sensitivity, Zhao [26] proposed a dual-mode thin film bulk acoustic resonator (DM-FBAR) temperature sensor that uses different phosphorus-doped silica insertion layers. The FBAR with a SiO_2 insertion layer doped with 4 sccm PH3 achieved a high-temperature sensitivity of up to 64.8 kHz/°C (resonant frequency magnitude of GHz).

3.4 Micro/nano temperature sensor

Micro/nano temperature sensors are typically fabricated using special treatment of materials on a micron or nano scale. Common materials used for fabrication include metals, semiconductors, polymers, and nanoparticles. Due to their small size, high sensitivity, fast response time, and ability to achieve local measurements, they are widely used for temperature monitoring in microelectronic devices, biomedical research, the automotive industry, and other fields.

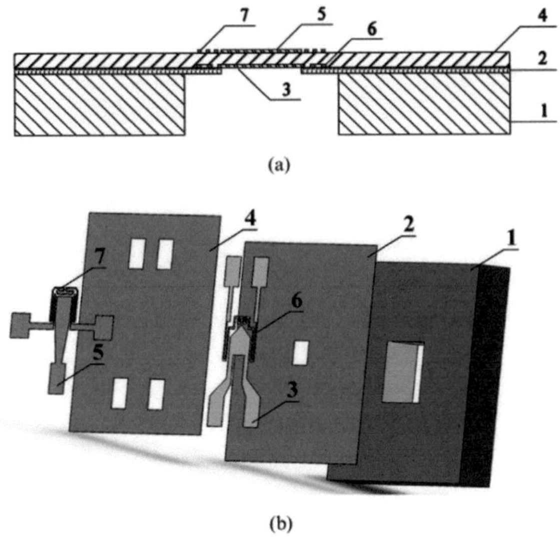

Figure 15.
Structure of the designed FBAR sensor chip: (a) cross section and (b) exploded view. 1-Si substrate, 2-silicon nitride support film, 3-bottom electrode, 4-ZnO piezoelectric film, 5-top electrode, 6-resistor heater, and 7-resistor temperature sensor [25].

3.4.1 Micro/nanoscale thermocouple

Micro/nano thermocouples use the same measurement principle as traditional thermocouples, which is based on the thermoelectric effect of materials for temperature measurement. The difference lies in the use of modern micro or nano processing technology to produce a micro and nanoscale probe as a measurement tool. This allows for the measurement of temperature at a very small scale and with high sensitivity, making them useful in various fields including microelectronics, biomedicine, and nanotechnology.

Micro/nano thermocouples can be prepared using electrochemical etching techniques to create a micro/nanoscale probe. However, a new method was proposed by Huang [27] that involves depositing carbon and platinum metals on the inner and outer surfaces of quartz nanopipettes such as **Figure 16** shown. This method achieved an average temperature sensitivity of 1.98 ± 0.07 μV/K during calibration in the normal temperature range (−10–20°C), with a temperature resolution of 0.08–0.24°C. While the structural stability of this thermocouple preparation method has yet to be verified, Gu's team [28] successfully prepared micro and nanoscale thermocouples (W-Pt, 100 nm probe) with good structural stability. They obtained a temperature resolution of less than 0.1°C and a temperature response time of about 400 ns and demonstrated its effectiveness in measuring cell temperature in subsequent experiments [29]. Due to their stability and smaller thermal inertia compared to traditional thermoelectric devices, micro/nano thermocouples have great potential for applications in microelectronics and microbiological industries.

3.4.2 Magnetic nanoparticles

Magnetic nanoparticles possess unique properties that distinguish them from traditional magnetic materials, mainly due to their nanometer scale. These properties

DOI: http://dx.doi.org/10.5772/intechopen.112877

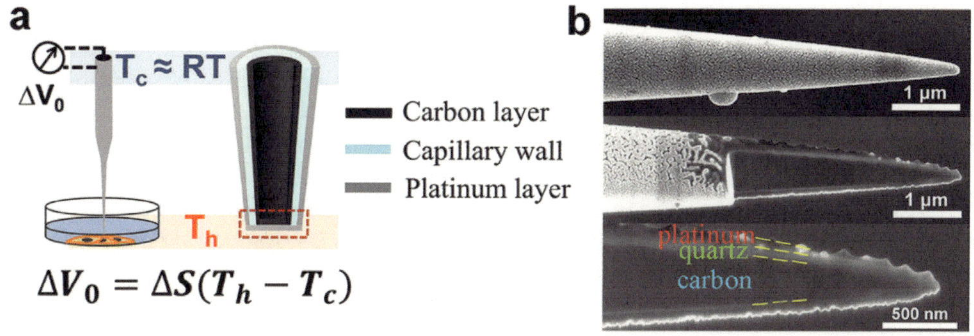

Figure 16.
(a) Schematic diagram of a nano thermocouple probe for intracellular temperature sensing and (b) Nano thermocouple probe [27].

include magnetic transparency and superparamagnetic effects. The former characteristic is particularly advantageous for minimizing potential harm to biological organisms during temperature measurement. Additionally, according to Langevin's magnetic model [30], the magnetization of superparamagnetic particles is temperature-dependent. Thus, the theoretical foundation for magnetic nanoparticle temperature measurement (MNT) has been established, and a corresponding measurement device schematic is presented in **Figure 17**.

Xu's group [32] has developed a temperature measurement technique utilizing magnetic nanoparticles (MNPs) based on a real-time DC magnetization model. The MNP temperature-magnetic response rate was analyzed using theories related to heat transfer and ultrafast magnetic dynamics, leading to an innovative approach that utilizes both the frequency-domain response and time-domain response of MNPs. The frequency-domain response is utilized for parameter calibration, while the time-domain response is utilized for solving the temperature variation, thus achieving semi-invasive transient temperature measurement. Direct heating of the MNP with single pulses of varying pulse widths resulted in a rapid temporal resolution of 14.4 ns. In addition to its fast response time, this method also boasts a satisfactory

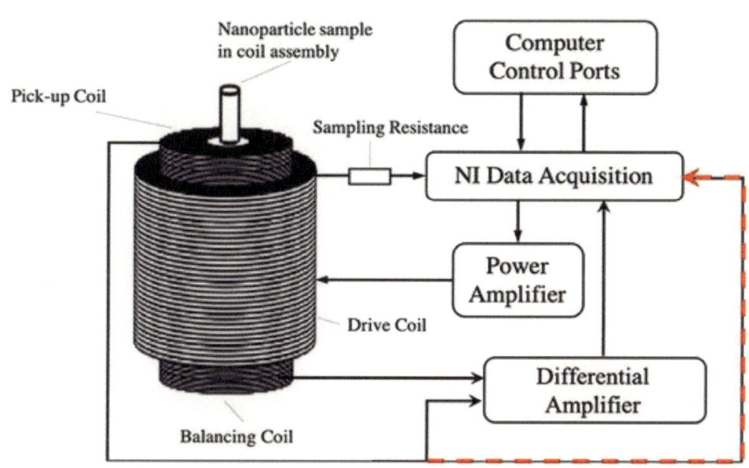

Figure 17.
MNT process diagram [31].

temperature resolution. Meanwhile, Wang [33] enriched the harmonic signal by applying a dual-frequency magnetic field, reducing the thermal noise level through the principle that mixed-frequency harmonic information can reduce the sampling bandwidth [31], and detecting the magnetization signal with a tunnel magnetoresistance (TMR) element to obtain the analog signal of temperature. The four-harmonic model allows the temperature measurement error to be controlled within 0.015K.

Given the high accuracy of measurement, small particle size, and excellent temporal resolution associated with magnetic nanoparticles, magnetic particle imaging (MPI) [34] is a highly promising technique for a wide range of applications in industry and biomedicine, including the precise and noninvasive measurement of temperature.

3.4.3 Micro/nanoscale fluorescent particles

Fluorescent particle micro-clusters [35], such as organic dyes, lanthanide chelates, quantum dots, and lanthanide-doped nanoparticles, utilize the thermal properties of the material to emit light with different intensities, spectral distributions, and decay lifetimes at different internal electron energy levels and under different temperature conditions.

Erving C [36] injected a solution containing fluorescent nanoparticles (NPs) into a mouse and placed them in an imaging chamber at a constant temperature environment. Two wavelengths of laser light were used to achieve tissue heating (808 nm) and optical excitation of the particles, respectively. The results were recorded by a camera and the fluorescence intensity was converted to temperature using a software algorithm. This measurement method effectively demonstrated the phenomenon of local heating in mice, with a temperature resolution of 0.6°C, as expected (**Figure 18**).

This technology offers a combination of high thermal sensitivity (>1%/K) and spatial resolution (<10 μm). It has a short collection time (<1 ms) and can be operated

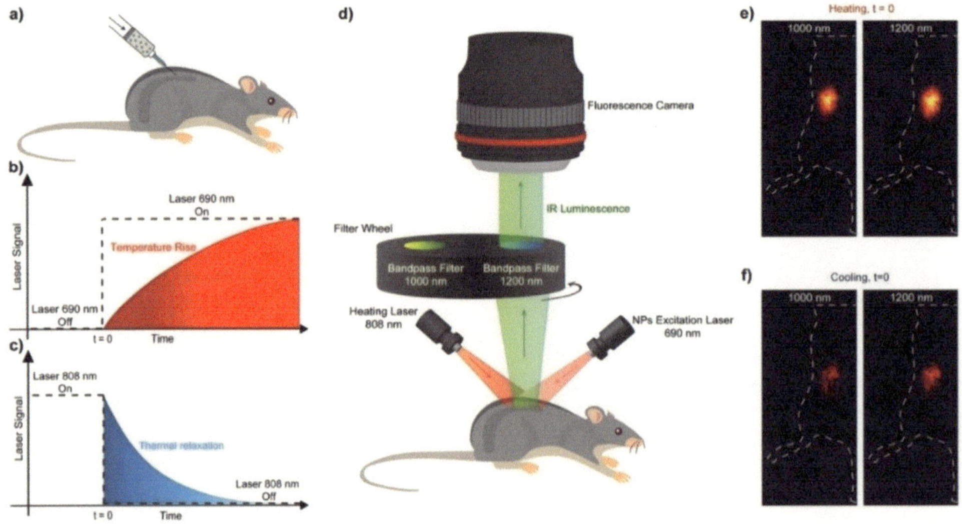

Figure 18.
(a) Subcutaneous injection of thermal sensitivity NPs in a mouse; temperature profiles as b) heating and c) cooling take place; (d) schematic representation of the in vivo SDTI experiment; fluorescence images of (e) heating and (f) cooling process [36].

remotely, making it suitable for use in biological fluids, with rapidly moving objects, and in strong electromagnetic fields. Due to its real-time observation capabilities and intuitive visualization, fluorescent particle clusters have unique advantages in monitoring the process of cellular life activities.

4. Conclusion

Traditional temperature sensors are widely used in various fields due to their simple structure, low manufacturing cost, and high stability. However, they face limitations in complex and harsh environments, which are prevalent today. This paper highlights several types of advanced temperature sensors that offer unique advantages and applications compared to traditional sensors. Different temperature sensors are used for varying extreme application scenarios such as acoustic temperature sensors for high-temperature boilers, fiber optic temperature sensors for long-distance cable lines, nondestructive temperature measurement of human tissues, and micro and nanoscale temperature sensors for microstructure environments. Most of the new temperature sensors feature contactless measurement and remote control.

In addition, based on the complexity of multi-temperature measurement, temperature resolution is no longer the only criterion for judging the performance of temperature measurement. For different application scenarios, people pursue different advantages of temperature sensors, spatial resolution, signal-noise rate, and response time, which have become the key information we also need to focus on in today's temperature measurement process.

Author details

Deqi Chen[1*], Qianlong Zuo[1], Hao Wu[2], Haidong Liu[1] and Fenglei Niu[3]

1 Department of Power and Energy Engineering, Chongqing University, Chongqing, China

2 School of Nuclear Science and Engineering, North China Electric Power University, Beijing, China

3 Microfluidic Foundry L.L.C., San Pablo, CA, USA

*Address all correspondence to: chendeqi@cqu.edu.cn

References

[1] Federation CMI. Industrial Thermocouple Assemblies: GB/T 30429-2013. Beijing, China: Standards Press of China; 2013

[2] Li J, Tao B, Huang S, et al. Cutting tools embedded with thin film thermocouples vertically to the rake face for temperature measurement. Sensors and Actuators A Physical. 2019;**2019**:296

[3] Federation CMI. Industrial Platinum Resistance Thermometers and Platinum Temperature Sensors: GB/T 30121-2013. Beijing, China: Standards Press of China; 2013

[4] Mbarek SB, Alcheikh N, Younis MI. Recent advances on MEMS based infrared thermopile detectors. Microsystem Technologies. 2022;**28**(8):1751-1764

[5] Shen C. Performance enhance of CMOS-MEMS thermoelectric infrared sensor by using sensing material and structure design. Journal of Micromechanics and Microengineering. 2019;**29**(2):025007

[6] Hou H, Huang Q, Liu J, et al. Si_3N_4-TiN loaded carbon coating with porous structure as broadband light superabsorber for uncooled IR sensors. Infrared Physics & Technology. 2020;**105**:103240

[7] Li W, Ni Z, Wang J, et al. A front-side microfabricated tiny-size thermopile infrared detector with high sensitivity and fast response. IEEE Transactions on Electron Devices. 2019;**5**:1-8

[8] Edwards GJ, Levick AP, Xie Z. Laser emissivity free thermometry (left). In: TEMPMEKO 96, 6th International Symposium on Temperature and Thermal Measurements in Industry and Science, 10-12 September 1996, Torino, Italy. 1996

[9] An BL, Qu Y, Song XY, et al. On surface temperature measurement of low emittance artefact coating by active infrared laser radiation thermometry. Infrared Physics & Technology. 2021;**115**(29):103696

[10] Wang A, Gollapudi S, Murphy KA, et al. Sapphire-fiber-based intrinsic Fabry–Perot interferometer. Optics Letters. 1992;**17**(14):1021-1023

[11] Yutong Z, Yi J, Xinxing F, et al. Review of sapphire Fiber high temperature Fabry-Perot sensor. Semiconductor Optoelectronics. 2022;**43**(4):10

[12] Cui Y, Jiang Y, Zhang Y, et al. An all-sapphire fiber temperature sensor for high-temperature measurement. Measurement Science and Technology. 2022;**33**(10):105115

[13] Feng X, Jiang Y, Xie S, et al. Higher-order mode suppression technique for multimode sapphire fiber external Fabry-Perot interferometers. Optics Express. 2022;**30**(4):4759-4767

[14] Feng X, Jiang Y, Xie S, et al. Signal pick-up technique for sapphire fiber external Fabry-Perot interferometer using tapered multimode fibers. Journal of Lightwave Technology. 2022;**40**(12):3992-3996

[15] Polz L, Dutz FJ, Maier RRJ, et al. Regenerated fibre Bragg gratings: A critical assessment of more than 20 years of investigations. Optics & Laser Technology. 2021;**134**:106650

[16] Gunawardena DS et al. Resurgent regenerated fiber Bragg gratings and thermal annealing techniques for ultra-high temperature sensing beyond 1400° C. Optics Express. 2020;**28**(7):10595-10608

[17] Dutz FJ, Heinrich A, Bank R, et al. Fiber-optic multipoint sensor system with low drift for the long-term monitoring of high-temperature distributions in chemical reactors. Sensors. 2019;**19**(24):5476

[18] Laarossi I, Quintela-Incera MÁ, López-Higuera JM. Comparative experimental study of a high-temperature Raman-based distributed optical fiber sensor with different special fibers. Sensors. 2019;**19**(3):574

[19] Silva M, Barros T, Alves HP, et al. Evaluation of fiber optic Raman scattering distributed temperature sensor Between-196 and 400 degrees C. IEEE Sensors Journal. 2021;**21**:2

[20] Bo L, Yu Z, Hill C, et al. Sapphire-fiber-based distributed high-temperature sensing system. Optics Letters. 2016;**41**(18):4405-4408

[21] Liu B, Buric MP, Chorpening BT, et al. Design and implementation of distributed ultra-high temperature sensing system with a single crystal fiber. Journal of Lightwave Technology. 2018;**36**(23):5511-5520

[22] Wang Y, Zou F, Cegla FB. Acoustic waveguides: An attractive alternative for accurate and robust contact thermometry – ScienceDirect. Sensors and Actuators A: Physical. 2018;**270**:84-88

[23] Jia R, Xiong Q, Wang L, et al. Study of ultrasonic thermometry based on ultrasonic time-of-flight measurement. AIP Advances. 2016;**6**(3):23-33

[24] Wang H, Zhou X, Yang Q, et al. A reconstruction method of boiler furnace temperature distribution based on acoustic measurement. IEEE Transactions on Instrumentation and Measurement. 2021;**70**:9600413

[25] Zhang M, Zhao Z, Du L, et al. A film bulk acoustic resonator-based high-performance pressure sensor integrated with temperature control system. Journal of Micromechanics and Microengineering. 2017;**27**(4):045004

[26] Zhao JH, Xing YH, Han J, et al. The research of dual-mode film bulk acoustic resonator for enhancing temperature sensitivity. Semiconductor Science and Technology. 2021;**36**(2):025018

[27] Huang LQ, Ding XL, Pan XT, et al. Single-cell thermometry with a nanothermocouple probe. Chemical Communications. 2023;**59**(7):876-879

[28] Yang S, He W, Li C, et al. A new approach of electrochemical etching fabrication based on drop-off delay control. Review of Scientific Instruments. 2019;**90**(7):074902

[29] Bai T, Gu N. Micro/nanoscale thermometry for cellular thermal sensing. Small. 2016;**12**:4590-4610

[30] Kaiser R, Miskolczy G. Magnetic properties of stable dispersions of subdomain magnetite particles. Journal of Applied Physics, 1970, 41(3): 1064-1072

[31] Guo S, Liu J, Du Z, et al. Improving magnetic nanothermometry accuracy through mixing-frequency excitation. Review of Scientific Instruments. 2021;**92**(2):024901

[32] Xu W, Liu W, Zhang P. Nanosecond-resolved temperature measurements using magnetic nanoparticles. Review of Scientific Instruments. 2016;**87**(5):324-337

[33] Wang S, Zhang P, Liu W, et al. High-resolution magnetic nanoparticle temperature measurement method based on dual-frequency magnetic field excitation. Measurement Science and Technology. 2021;**32**(7):075701

[34] Gleich B, Weizenecker J. Tomographic imaging using the nonlinear response of magnetic particles. Nature. 2005;**435**(7046):1214-1217

[35] Zheng W et al. The development and future of temperature measurement for biosome and cells in Micro- nano meter scale. Acta Metrologica Sinica. 2022;**6**:043

[36] Ximendes EC, Rocha U, Sales TO, et al. In vivo subcutaneous thermal video recording by supersensitive infrared nanothermometers. Advanced Functional Materials. 2017;**27**(38):1702249

Chapter 8

Types of Temperature Sensors

Reuben S. Diarah, Christian Osueke, Adefemi Adekunle, Segun Adebayo, Adedayo Banji Aaron and Olaluyi Olawale Joshua

Abstract

There are three main types of temperature sensors: thermometers, resistance temperature detectors and thermocouples. These sensors measure a physical property that changes as a function of temperature, and temperature sensors are classified into contact and non-contact sensors. Contact sensors detect the degree of hotness or coldness of an object when placed in direct contact with the object. It can be used to sense the degree of hotness or coldness in liquids, solids or gases in a wide range of temperatures. Contact temperature sensors include thermometers, thermocouples and thermistors. A thermometer detects the body temperature of human beings, and a thermocouple is a thermoelectrical thermometer that works on the principle of the Seebeck effect; they are cheap; hence, their model and basic materials are easy to get, and non-contact sensors are not placed in contact with the object that it measures; however, they measure the temperature by utilizing the radiation of the heat source. IR sensors detect the energy of an object remotely and emit a sign to an electronic circuit that senses the object's temperature by a specific calibration diagram. Other types of temperature sensors are available and produced based on the working principle, size, temperature range and their function and application.

Keywords: sensors, temperature, thermometer, thermistor, non-contact type sensor, contact type sensor

1. Introduction

A temperature sensor is an electronic device that measures the temperature of its environment and converts the input data into electronic data to record, monitor or communicate temperature changes. A temperature sensor is an electronic device that monitors the temperature of its surroundings and turns the input data into electronic data. Temperature sensors come in a wide variety of forms [1].

Temperature sensors are electrical/electronic physical sensing device which transforms an input signal from a specific environment into an equivalent output signal [2].

According to the amount of general literature on the topic, thermocouples are the most often employed type of temperature measuring in industry. Its widespread acceptance, reasonable accuracy over a wide measurement range, and relatively inexpensive sensors all contribute to its appeal. Narrower measuring ranges can handle accuracy closer to 0.1 degrees Celsius, whereas accuracy over wide ranges is comfortably between 0.5 and 2 degrees Celsius [3].

As long as the Seebeck coefficients of material A and material B for the two materials are known, these thermoelectric devices use the Seebeck effect in dissimilar metal wires linked at the thermoelectric junction representing T1 to determine a temperature gradient down the wire [4]. The temperature can be gauged at the terminus connections T0 by measuring the net electromotive force between T0 and T1 within the wires, which is voltage of the order of microvolts. Cold junctions are frequently utilised in the form of a fixed physical temperature or electronically mimicked via cold junction compensation because a temperature gradient must be constructed to produce a net voltage output signal (CJC).

Due to the non-linear temperature-resistance connection of thermistors, which are composed of semiconductor materials, calibration is even more crucial [5]. Although routine calibration is required to prevent the impacts of sensor drift, the use of semiconductor materials allows them to deliver a far better level of sensitivity [6] than other sensor types.

According to Schweiger's 2007 argument, if the right sensors are chosen and calibrated properly, a quick multichannel precision thermometer might compete with Precision Thermometers using thermistors [7]. Deviations of less than 30 mK were seen in tests conducted in the temperature range of −50 to 10 C. Improvements have been made in spatial resolution of surface temperature measurement compared to standard soldered type K thermocouple using an electrochemically etched microtip [8]. Thin film thermocouples can also be deposited onto a surface and have been used to measure heat generated in the friction between sliding surfaces [9]. Non-linearity

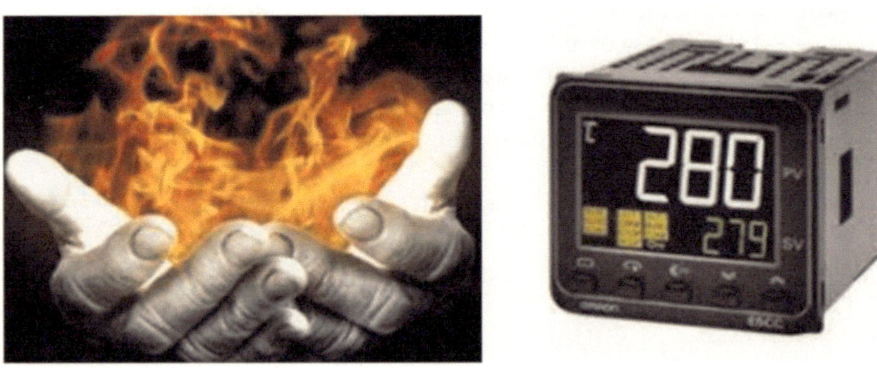

Figure 1.
Temperature-Sensing illustration [12–14].

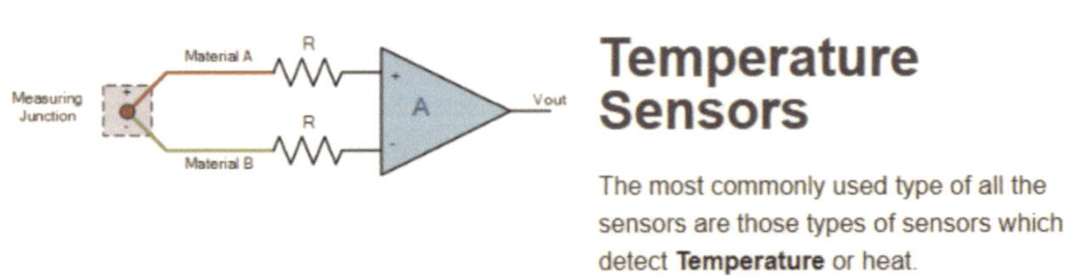

Figure 2.
Temperature sensor [15].

of sensors can be an issue, although one study showed it to be possible to correct for this using a neural network approach in type K thermocouples [10].

Industrial thermocouple measurements can be further enhanced by improving high-temperature alloys and more intelligent electronics [11].

Figure 1 shows an illustration of temperature-sensing using human hands as a sensor and its digital equivalent, while **Figure 2** shows a temperature sensor formed by joining two different materials. There are many different types, sizes and shapes of temperature sensors. In general, temperature sensors can be categorised into two groups: contact sensors and non-contact sensors [15].

2. Contact sensors

When positioned close to an object to be detected for heat or cold, contact sensors are used to measure the object's temperature. These sensors can determine the concentration of liquids, solids or gases throughout a wide temperature range.

Thermocouples and thermistors are good examples of contact temperature sensors.

Thermocouples are inexpensive, and it is easy to find the basic materials needed to manufacture thermocouples [15, 16].

Contact Sensors are devices that measure temperature by placing it in direct contact with the object being measured or the desired measurement environment. They can be used to detect temperature changes in gases, liquids or solids in a range of temperature measurements. Thermocouples and thermistors are two contact sensor types. Its model and fundamental components are straightforward, and thermocouples are frequently inexpensive.

Additionally, thermocouples have the broadest temperature range of any temperature sensor, ranging from well below -200°C to well over 2000°C [16].

Thermocouples are thermoelectric sensors that are essentially made of two welded or crimped junctions of dissimilar metals, such as copper and constantan. The reference (cold) junction and the measuring (hot) junction are the two junctions that are maintained at the same temperature. As illustrated below, a voltage is created across the junction when the two junctions are at different temperatures. This voltage is used to measure the temperature sensor [16].

2.1 Construction of a thermocouple

Figure 3 shows how a thermocouple is constructed by joining two metals of iron and constantan.

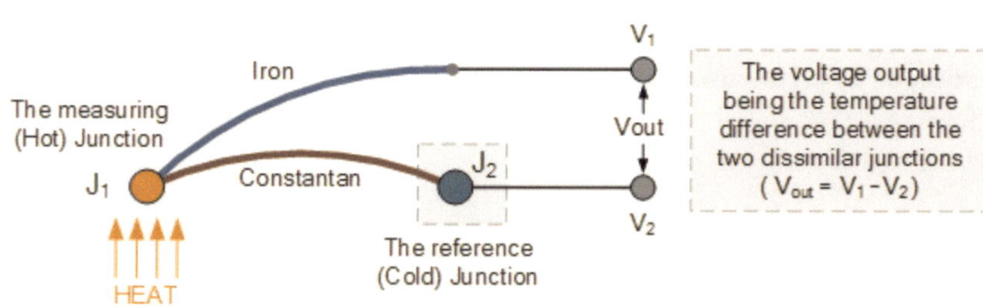

Figure 3.
Construction of a thermocouple [16].

2.2 Working principle of a thermocouple

The thermocouple's working principle is quite straightforward and fundamental. When two different metals, such as copper and constantan, are fused together, a "thermoelectric" effect results, producing a constant potential difference between the two materials of only a few millivolts (mV). The "Seebeck effect" refers to the voltage differential between the two junctions because an electromagnetic field (emf) is created when a temperature gradient develops between the conducting wires. The output voltage of a thermocouple is then dependent on temperature variations [17].

If both junctions in **Figure 3** are at the same temperature (zero potential difference across the junctions), and there is no voltage output because V1 = V2. But when the junctions are linked together in a circuit and operate at different temperatures, a voltage output, V1 - V2, corresponding to the temperature differential between the two junctions, will be noticed. This is because the characteristics of the two different metals employed influence how much of a voltage difference will increase with temperature until the junction reaches its maximum voltage level [17].

Extreme temperatures between −200°C and over +2000°C can be recorded using thermocouples, which can be constructed from various materials. Internationally recognised standards have been created with thermocouple colour codes to help users select the best thermocouple sensor for a given application due to the wide variety of materials and temperature ranges available. Below is a list of the standard thermocouple colours used in Britain [17].

Figure 4. shows the thermocouple colour codes that were used in the manufacturing of different types of thermocouples. Thermistor contacts are the second kind of contact temperature sensor. The resistance of thermistors is dependent on temperature change, as opposed to other types of resistors whose value is determined by the colour code [18].

Thermistors are available in two types which are:

1. Positive temperature coefficient (PTC)

2. Negative temperature coefficient (NTC)

A PTC thermistor's resistance rises with temperature, but an NTC thermistor's resistance falls with temperature. Therefore, an NTC thermistor is the most common type of thermistor.

Temperature sensors include thermocouples. They can be found in common appliances, including ovens, refrigerators and fire alarms. Thermometers and numerous other vehicle appliances also include them [18].

Figures 5 and **6** show PTC (left) and NTC (right) thermistor electrical symbols and a typical NTC thermistor.

2.3 Advantages of a thermistor

- Less expensive

- Can measure changes in a small temperature range

- They are more sensitive than other temperature sensors

- They provide a fast response

Thermocouple Sensor Colour Codes*Extension and Compensating Leads*			
Code Type	Conductors (+/-)	Sensitivity	British BS 1843:1952
E	Nickel Chromium / Constantan	-200 to 900°C	
J	Iron / Constantan	0 to 750°C	
K	Nickel Chromium / Nickel Aluminium	-200 to 1250°C	
N	Nicrosil / Nisil	0 to 1250°C	
T	Copper / Constantan	-200 to 350°C	
U	Copper / Copper Nickel Compensating for "S" and "R"	0 to 1450°C	

Figure 4.
Thermocouple colour codes [17].

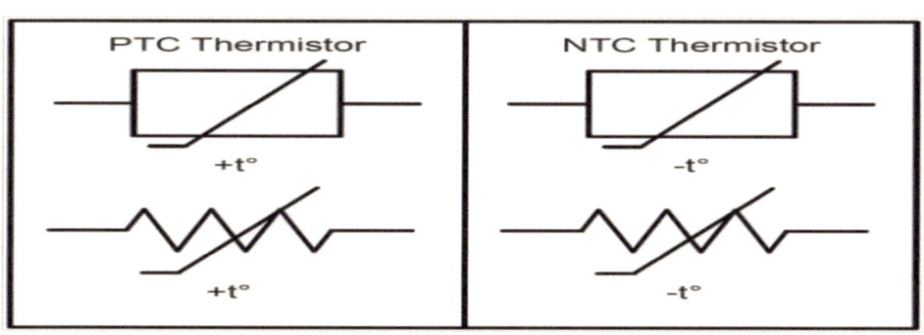

Figure 5.
PTC (left) and NTC (right) thermistor electrical symbols [19].

- They are easy to use

They are small and can fit into any smallest space [19].

A bi-metallic strip is created when two distinct metals, such as nickel, copper, tungsten, or aluminium, are bonded together to create the thermostat, an electro-mechanical

Figure 6.
A typical thermistor [19].

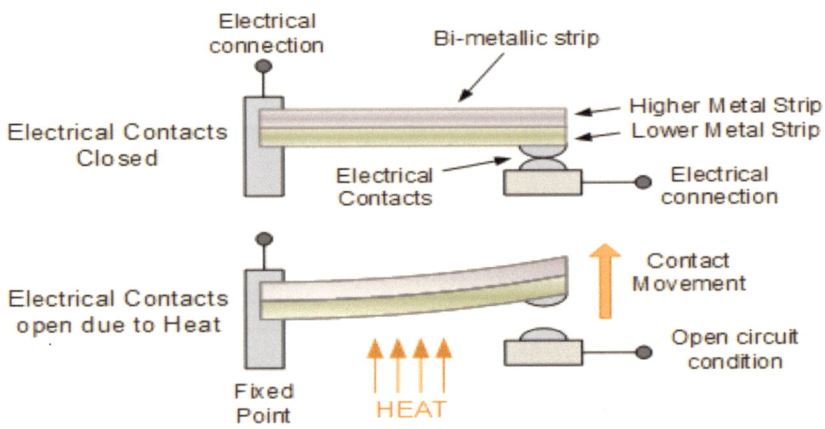

Figure 7.
Bi-metallic strip.

contact type temperature sensor. When the strip is heated, the differing linear expansion rates of the two dissimilar metals cause a mechanical bending action.

The bi-metallic strip is frequently used to control hot water heating elements in boilers, furnaces, hot water storage tanks and vehicle radiator cooling systems. In addition, it can be used as an electrical switch on its own or as a mechanical method of operating an electrical switch in thermostatic controls [16].

Figure 7 shows two metals with distinct thermal properties bonded back-to-back to form the thermostat. The connections are closed when it is cold, allowing current to flow through the thermostat. However, the bonded bi-metallic strip bends up (or down) and opens the contacts when it gets hot because one metal expands more than the other, blocking the current flow [16].

2.4 Thermostat

A thermostat is a temperature-sensing tool that gauges engine coolant temperature. In order for internal combustion engines to operate at an efficient temperature, the component is intended to know when to open and close.

Figure 8.
On/off thermostat [16].

If the coolant is not hot enough, the thermostats stay closed. However, when the coolant reaches a certain temperature, a valve opens, letting hot coolant flow into the radiator. The thermostat therefore functions similarly to a gate by allowing or preventing the passage of coolant from the engine to the radiator.

Modern automobile engines operate within a specific temperature range; typically, they operate between 194 degrees Fahrenheit, or 90 degrees Celsius, and 221 degrees Fahrenheit. The thermostat determines when to open and close based on the coolant temperature [20].

Figure 8 shows the on/off the thermostat; there are two main types of bi-metallic strips with respect to their movement when subjected to temperature changes. They are:

1. snap-action

2. creeper types

Both the faster "creep-action" types gradually adjust their position as the temperature changes, and the snap-action types generate an instantaneous "ON/OFF" or "OFF/ON" type action on the electrical connections.

Snap-action type thermostats are frequently used in our houses to regulate the temperature set point of ovens, irons, immersion hot water tanks, as well as the domestic heating system. They can also be found mounted on walls [16].

In most creeper varieties, a bi-metallic coil or spiral slowly unwinds or coils up in response to temperature changes. Since the creeper-type bi-metallic strips are longer and thinner than the conventional snap ON/OFF varieties, they are typically more sensitive to temperature changes, making them perfect for use in temperature gauges, dials, and other similar devices [16].

Standard snap-action-type thermostats have a significant hysteresis range between the time the electrical contacts open and the time they close again, which is a drawback despite their low-cost and wide operating range when used as temperature sensors. It might be set to 20°C, for instance, but not open until 22°C or close again until 18°C [16].

Therefore, the temperature swing range might be rather wide. Bi-metallic thermostats that are sold for residential usage contain temperature adjustment screws that enable more exact pre-setting of the appropriate temperature set point and hysteresis level [16].

Contact sensors are employed in industries to control various automation temperature processes; hence, it is advantageous to use sensors in the industry, offices and home to regulate the environment's temperature.

2.5 What is a temperature controller?

Temperature controls make sure a process gets the desired temperature and keeps it there. These are typically employed for closed-loop control, in which the temperature controller compares the actual temperature with the set point established by the programmer using data from a temperature probe (thermocouple, resistance thermometer or temperature transmitter). It then modifies its output signal to the appropriate control element as necessary (electrical heater, cooling circuit, steam control valve, etc.). A variable output, where the output signal to the process is between 0 and 100%, and a straightforward ON/OFF control, working like a thermostat, are possible. The latter is also called a 2-point, binary, or bang-bang control [21].

2.6 How does a temperature controller work

The heating circuit is turned on for ON/OFF control when the temperature is below the set point and off when it is above. Additionally, a cooling circuit may be activated above and deactivated below the specified point. A proportional–integral–derivative (PID) controller frequently performs variable control (three-term controller). In order to attain and keep the set point with the least amount of overshoot and to retain it as steadily as possible, this controller applies a revised algorithm on the error (the difference between the set point and the measured value) [21].

2.7 What is a PID controller

Depending on the needs of the process, three-term or PID controllers (proportional–integral–derivative) can be employed for proportional alone (P), PI or PID control. In proportion to the departure from the set point, proportional control modifies the output. A defined proportionate band is below and/or above the set point. The output for cooling (above) or heating (below) is 100% outside of this band. It decreases linearly within the band, reaching 0% at the set point. The integral term can then further alter the output based on the rate-of-change of the mistake because this can result in a sluggish approach to the set point (achieving the set point quicker). Due to the possibility of overshooting the fixed point, the derivative term predicts future errors and modifies the output [21].

2.8 Advantages of temperature sensors

Temperature sensors are possible when an object needs to be heated, cooled, or both, and it must maintain the desired temperature (setpoint) despite changes in its surroundings.

Open-loop and closed-loop controls are the two fundamental methods of temperature control.

Open-loop systems apply continuous heating and cooling without considering the actual temperature output. It is comparable to a car's interior heating system. You might have to set the heat all the way up on a chilly day to get the car up to 75 degrees. However, during warmer weather, the same setting would leave the inside of the car much warmer than the desired 75° [22].

Temperature sensors can control a given situation using the open and closed loops, as shown in **Figures 9** and **10**, respectively.

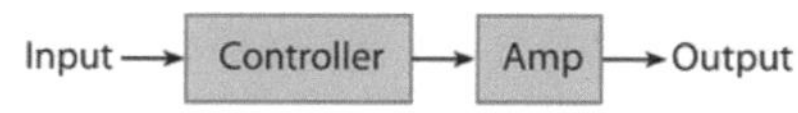

Figure 9.
Open-loop temperature control diagram [23].

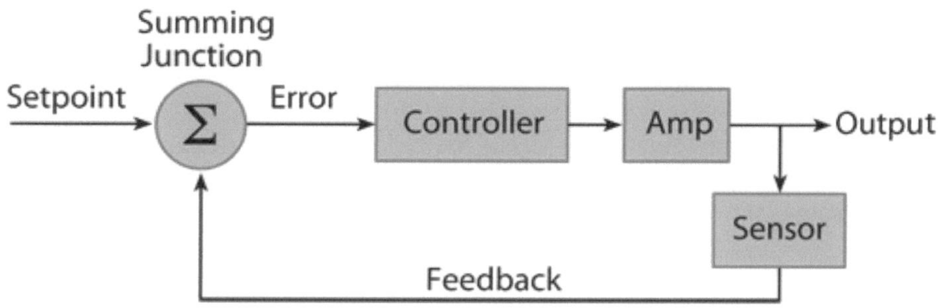

Figure 10.
Close loop temperature controller block diagram [13, 23, 24].

Regardless of sophistication, all temperature sensors and controllers operate in essentially the same way. A controller keeps a variable or parameter constant at a predetermined value. The actual input signal and the desired setpoint value are the two variables that the controller needs. The input signal is also known as the process value. The controller determines how frequently the input is sampled [25].

The input or process value is then compared to the setpoint value. If the process value deviates from the setpoint, the controller changes the output signal based on the difference between the process value and the setpoint and whether the process value is getting closer to the setpoint or moving further away from it. The actual value is then changed in response to the output signal in order to bring it into compliance with the setpoint. Typically, the control algorithm updates the output power value before applying it to the output [25].

The control action is based on the type of controller being used. The controller decides whether the output should be turned on, off or left in its current state, for example, if it is an ON/OFF control [25].

One of the easiest control kinds to use is the ON/OFF control. By establishing a hysteresis band, it operates. To regulate the temperature inside a room, for instance, a temperature controller might be used. An error signal would display a $-1°$ difference if the setpoint temperature was 68° and the actual temperature was 67°. The temperature would then be raised back to the setpoint of 68° by the controller sending a signal to increase the applied heat. The heater turns off when the room reaches 68 degrees. The controller does nothing, and the heater stays off for a temperature between 68° and 67°. The heater will, however, start up once the temperature hits 67° [25].

Unlike ON/OFF control, PID control determines the precise output value required to maintain the desired temperature. Power output ranges from 0–100%. When an analogue output type is used, the output drive is proportional to the output power value. If the output is a binary output type, such as a relay, Solid State Relay driver or triac, it must be time-proportional in order to provide an analogue representation [25].

A system that uses cycle time to proportion output values is called time-proportional. A system requiring 50% power will have its output on for 4 seconds and off for 4 seconds if the cycle time is set to 8 seconds. The time values would not change as long as the power value remained constant. The power is gradually averaged to the requested 50% amount, which is evenly split between on and off. The output would be on for two seconds and off for six seconds over an eight-second cycle if the output power needed to be 25% [25] as shown in **Figure 11**.

A shorter cycle time is desired, barring any other factors, because the controller can react to changes in the process and the output's condition more quickly. Due to the way relays operate, which may shorten their longevity, a cycle duration of less than 8 seconds is not recommended. For solid state switching components like an SSR driver or triac, quicker switching times are preferred. Longer switching times allow for higher process value variation regardless of the output type. A longer cycle time is typically desirable when employing a relay output, but only if the process allows it [25], as shown in **Figure 12**.

Table 1 shows the comparison between NTC thermistor and thermocouple.

2.9 Non-contact sensors

Non-contact sensors are not in contact with the object that it measures; however, they measure the temperature by utilising the radiation of the heat source. An example of a non-contact sensor is the infrared (IR) sensor. IRs detect the energy of an object remotely and emit a sign to an electronic circuit that senses the object's temperature by a specific calibration.

Non-contact temperature sensors generally rely on technologies that are based on electrical, magnetic, optical, sonic or other principles rather than depending on physical contact or mechanical movement to obtain the measurements. The sensor often

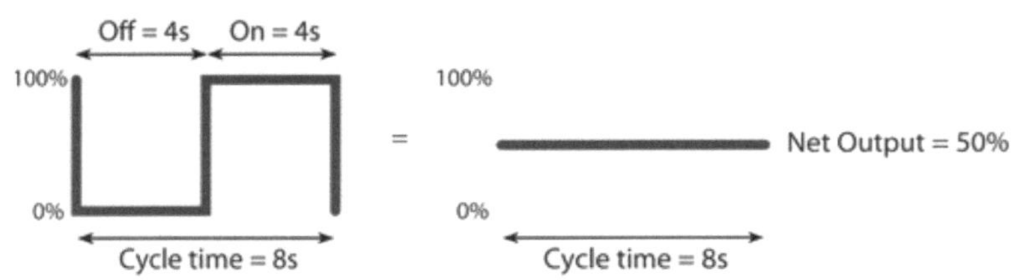

Figure 11.
Output time proportioning [25].

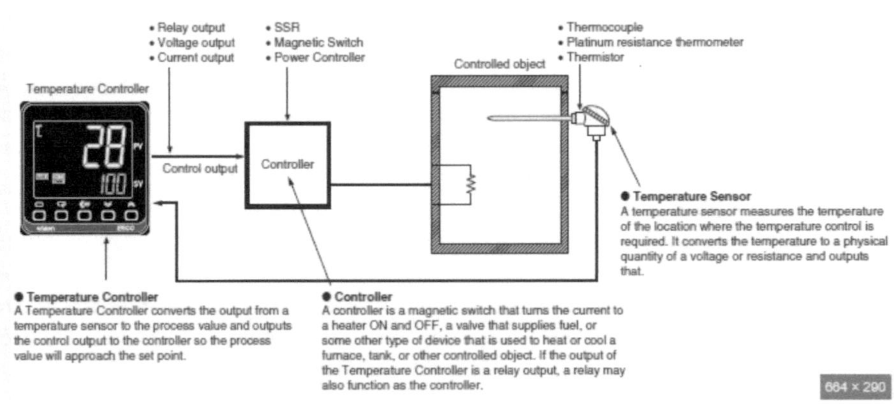

Figure 12.
Overview of contact temperature sensor controller [1].

NTC Thermistor	Thermocouple	
Effect of lead resistance on accuracy	Very low	None
Linearity	Non-linear-output requires linearisation	Non-linear requires conversion
Stability	Epoxy coated: 0.2 °C/year Hermetically sealed: 0.02°C/year	>1 °C/year
Response time	0.12–10 s (depending on size and packaging)	0.2–10 s (depending on size and packaging)
Temperature range	−50 to 250°C (dependent on type)	−200 to 1250°C, dependent on type
Temperature range	−50 to 250°C (dependent on type)	−200 to 1250°C, dependent on type
Response time	0.12–10 s (depending on size and packaging)	0.2–10 s (depending on size and packaging)
Stability	Epoxy coated: 0.2°C/year Hermetically Sealed: 0.02°C/year	>1°C/year
Linearity	Non-linear-output requires linearisation	Non-linear requires conversion
Effect of lead resistance on accuracy	Very low	None

Table 1.
A brief comparison of thermistor and thermocouple [19].

emits a form of energy such as radiation that can be used to detect a condition without physical contact.

2.10 Working principle of non-contact sensors

Non-contact sensors detect changes in physical environmental conditions without physical contact with the measured object. There are several types of non-contact sensors, including optical, capacitive, magnetic, ultrasonic and many other types of sensors.

The specific working principle of non-contact sensors can vary based on the type of sensing pattern; however, they all depend on detecting changes in the environment, converting the required information into electrical signals which can be processed or analysed.

Manufacturing process of non-contact sensors.

Non-contact sensors are manufactured using a variety of different technologies depending on the specific application and the type of sensor being produced.

Optical sensors: Optical sensors use light to detect changes in position or distance. They can be made using a variety of.

2.11 Applications of non-contact sensors

The field is progressing thanks to innovation. Active-matrix flexible temperature sensors and self-powered flexible temperature sensors are two examples of flexible temperature sensors that have recently been studied and optimised. Flexible temperature sensors also include flexible thermocouples, flexible thermistors, and flexible thermochromic types [26].

Patients' temperatures have been monitored using printable, flexible sensors with excellent sensitivity. There is a trend toward creating wearable sensors that can measure temperature, avoiding conventional problems with heavy equipment and measuring inaccuracies caused by a variety of factors such as the wearer's movement.

Other prominent research in the field of non-contact infrared temperature sensors is recent work on creating a low-cost, more accurate Arduino-based infrared thermometer for body temperature detection. Arduino is an open-source electronics platform that converts input to output. This research aims to circumvent the problems inherent with non-contact infrared sensors currently on the market [26] (**Table 2**).

Advantages of non-contact temperature sensors.

1. They are used in measuring hard-to-reach or very hot objects.

2. They have very short measurement and response time.

3. They are used in the non-destructive measurement

4. They have longevity of measuring point.

5. They have the option of measuring even at high voltages, electromagnetic fields or aggressive materials.

Examples of non-contact temperature sensors

1. Thermal imagers

2. Furnace monitoring cameras

3. Infrared thermometers

4. Hall effect sensors technology

5. Ultrasonic sensors technology

6. Photonic sensors technology

7. Capacitive sensors technology

8. Inductive sensors technology

9. Laser displacement sensors technology

10. Radiation thermometers

11. Optical pyrometers

Type	Advantages	Disadvantages	Max working distance
MMW-Radar	1) Long working distance	1) Unapplicable for static objects	5 m–200 m
	2) Available for radial velocity	2) Generating false alarms easily	
	3) Applicable for all-weather		
Camera	1) Excellent discernibility	1) Heavy calculation burden	250 m (depending on the lens)
	2) Available lateral velocity	2) Light interference	
	3) Available for colour distribution	3) Weather susceptible	
		4) Unavailable for radial velocity	
LiDAR	1) Wide field of view (FOV)	1) Insufferable for bad weather	200 m
	2) High-range resolution	2) High price	
	3) High-angle resolution		
Ultrasonic	1) Inexpensive	1) Low resolution	2 m
		2) Inapplicable for high speed	
DSRC	1) Applicable for high speed(up to 150 km/h)	1) Low data rate	300-1000 m
	2) Relatively mature technology	2) Small coverage	
	3) Low latency (0.2 ms)		
LTE-V2X	1) Long working distance	1) High latency in long distance (> 1 s)	Up to 2 km
	2) Relatively high data transmission rate(Up to 300 Mbps)	2) Inapplicable for time-critical events	
5G-V2X	1) Ultra-high data transmission rate	1) Immature application	100 m - 300 m
	2) Low latency(< 80 ms)		
	3) High bandwidth		
	4) Applicable for high speed (up to 500 km/h)		

Table 2.
Comparison of different types of non-contact sensors.

Applications of temperature sensors.
Some temperature sensor applications include;

- Motorsport and other vehicles – within motorsports, there are many temperature sensor applications. These include; ensuring motors do not overheat, surface plate temperature, exhaust gas temperature, oil temperature, etc.

- Industrial equipment – most machinery used in manufacturing will contain a temperature sensor for safety reasons. Temperature sensors used within this environment must be highly robust and resistant to dirt and moisture.

- Medical Applications – temperature sensors are used for patient monitoring and within machines and devices for a range of medical procedures. In this industry, temperature sensors will require various safety standards and approvals.

- Food and beverage industry – temperature sensors are used within this environment as part of food safety standards, ensuring food is kept at the correct temperature. They are also used on various manufacturing equipment used within this sector.
- Home appliances and white goods – many appliances within the home will contain a temperature sensor, oven, toaster, kettles, washing machines, coffee machines, dishwashers, electric radiators, boilers, etc.
- Computers and devices – temperature sensors are used within computers and other devices to ensure they do not overheat and become dangerous.

More temperature sensor applications and areas:

- Calibration and Instrumentation
- Transit – refrigerated vans and lorries
- HVAC – Heating ventilation and air conditioning
- Power and utilities
- Renewable energy
- Heat Exchangers
- Drilling
- Laboratory and testing applications

3. Recent development in the temperature sensors

There is a recent development in temperature sensors, thanks to innovation. Active-matrix flexible temperature sensors and self-powered flexible temperature sensors are two examples of flexible temperature sensors that have recently been studied and optimised. Flexible temperature sensors also include flexible thermocouples, flexible thermistors, and flexible thermochromic types [26].

Printable, high-sensitivity flexible sensors have been explored to provide temperature monitoring of patients. There is a trend toward developing wearable sensors that can monitor temperature, circumventing traditional issues with bulky equipment and errors in measurement due to numerous factors such as the wearer's movement [26].

Gems sensors are made to detect or measure a media (air, gas, oil, water, steam, etc.). It may occasionally be essential to modify its attributes (level, volume, flow, pressure and temperature). Sometimes all that is required is to observe or record the media properties.

In order to represent the measurement or detection of the media, sensors send an output signal. After receiving the output, west controllers' devices can display, record, and/or control the process to modify the media's attributes to suit the application [27].

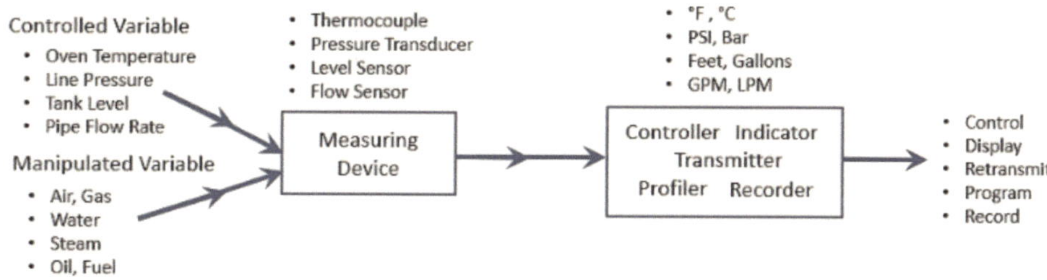

Figure 13.
Gems measurement continuous sensor [27].

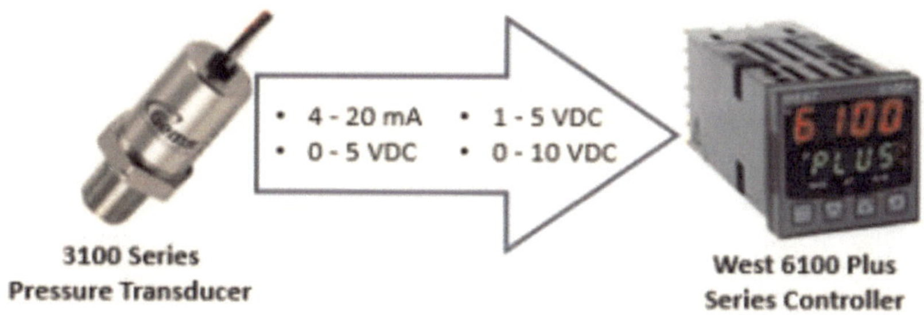

Figure 14.
The gems 3100 series pressure transducer.

Figure 13 shows gems continuous measurement sensors deliver a linear output to reflect the whole sensor range. DC voltage, current and frequency outputs are the three most typical linear outputs for sensors [27].

On the majority of versions, west controllers include a universal input. This supports most linear output kinds from gems sensors. However, it would help if you made sure the needed output is supported because west controllers do not support all output types offered by gems sensors [27].

Figure 14 shows the Gems 3100 Series Pressure Transducer, which may deliver 4–20 mA (milliamp) current or DC voltage outputs of 0–5, 1–5 and 0–10 VDC. The scale of these numbers corresponds to the pressure range that the transducer was designed. For example, the transducer would supply 4 mA at 0 PSIG (pounds per square inch, gauge) and 20 mA at 750 PSIG (pounds per square inch, gauge) if the 4–20 mA output for 0–750 PSIG was used, respectively. The universal input for 4–20 mA and the complete range of the 0–750 scale are both programmed into the West 6100 Plus Series Controller.

3.1 Common non-contact sensor technologies and real-world applications

There have been some common non-contact sensor technology developments in recent years, through a series of innovations, research and development.

3.1.1 Capacitive sensor technology

These non-contact sensor varieties track changes in capacitance to gather important details about the movement or location of a specific target. A capacitor can store energy in an electric field between two plates known as electrodes. This technology targets the other capacitor plate; the capacitance sensor is the first. The amplitude of the AC voltage, when a fixed frequency AC current is delivered, serves as a gauge of the separation between the sensor and target [28].

Position sensing and dynamic and thickness measuring are typical uses for capacitive sensor technology. In addition, on workstations, conveyors and robots, capacitive sensors can be utilised to detect parts and count and monitor liquid levels.

Everyday devices, such as digital audio players, smartphones and tablets, leverage capacitive sensing touchscreen as input devices. These sensors can also replace mechanical buttons [28].

3.1.2 Laser displacement sensor technology

The high accuracy of distance, position and displacement measurements of targets at long ranges are well suited for laser displacement sensors, also known as laser triangulation sensors.

These sensors are utilised for displacement measurement in a wide range of applications and sectors, from automated process control and research and development testing to Original Equipment Manufacturer integration, inventory management and more.

They are designed to measure and check the levels of liquid and bulk materials as well as the position, size, surface profile, vibrations, and sensing of technical items [28].

3.1.3 Inductive sensors technology

Inductive sensors employ magnetic fields produced in the coil to assess a target's motion or location.

When targets are conductive, one kind of inductive sensor technology uses Eddy currents.

This kind of sensor creates an alternating magnetic field by applying an alternating current to a coil.

The field causes currents—Eddy currents—in the target when it gets close to the sensor.

A secondary magnetic field is created by these currents and opposes the sensor's magnetic field.

The interaction can be gauged and utilised to calculate how far away the sensor is from the target.

Due to its resistance to grease, filth, dampness, magnetic interference fields and harsh industrial settings, Eddy current sensors are appropriate for use in places with limited access.

The measurement of internal combustion engine cylinder vibrations or sheet metal thickness in roller gaps is two instances of this technique in action [28].

Author details

Reuben S. Diarah[1*], Christian Osueke[1], Adefemi Adekunle[2], Segun Adebayo[1], Adedayo Banji Aaron[3] and Olaluyi Olawale Joshua[4]

1 Bowen University Iwo, Osun State, Nigeria

2 Federal University Oye Ekiti, Ekiti State, Nigeria

3 Landmark University Omu Aran, Kwara State, Nigeria

4 Bamidele Olumilua University of Science, Education and Technology, Ikere-Ekiti, Nigeria

*Address all correspondence to: diarah.samuel@bowen.edu.ng

References

[1] Available from: https://www.fiercee lectronics.com›sensors›what-a-temp

[2] Available from: https://cdn.kyklo.co/assets/W1siZiIsIjIwMjEvMDQvMDUv MTQvMjMvNTcvYjkyYmY2MmQtZD YzYS00MWU1LT kwMmUt NWExMTc 5YjBkYzNkL09NUk9OJTIwLSUyMEU 1R0MtNjAwJTIwc2VyaWVzJTIwLSUy MERhdGFzaGVldC5wZGYiXV0?sha= 4bd4677eb0b5ac99

[3] Childs PRN, Greenwood JR, Long CA. Review of temperature measurement. Review of Scientific Instruments. 2000; **71**:2959-2978

[4] Doebelin EO. Measurement Systems Application and Design. Fourth ed. United States: McGraw-Hill; 1990

[5] Childs PRN. 6 - Resistance temperature detectors. In: Childs PRN, editor. Practical Temperature Measurement. Butterworth-Heinemann: Oxford; 2001. pp. 145-193

[6] Ibrahim D. Chapter 5 - thermistor temperature sensors. In: Ibrahim D, editor. Microcontroller Based Temperature Monitoring and Control. Oxford: Newnes; 2002. pp. 107-127

[7] Schweiger HG, Multerer M, Gores HJ. Fast multichannel precision thermometer. Instrumentation and Measurement, IEEE Transactions on. 2007;**56**:2002-2009

[8] Genix M, Vairac P, Cretin B. Local temperature surface measurement with intrinsic thermocouple. International Journal of Thermal Sciences. 2009;**48**: 1679-1682

[9] Kennedy FE, Frusescu D, Li J. Thin film thermocouple arrays for sliding surface temperature measurement. Wear. 1997;**207**:46-54

[10] Jianwen T, Yong Z, Xiaojun T, and Junhua L. Nonlinearity correction of the thermocouple based on neural network. In: Intelligent Systems, 2009. GCIS '09. WRI Global Congress on. 2009. pp. 28-32

[11] Schuh W, Frost N. Improving industrial thermocouple temperature measurement. AIP Conference Proceedings. 2003;**684**:497-502

[12] OmRon PID/ON-OFFE55 Temperature Controller indiaMART

[13] Temperature Controller Basics Handbook

[14] Available from: https://www.ixthus.co.uk/news-media/blog-archive/tempe rature-sensor-applications

[15] Available from: https://www.electronics-tutorials.ws

[16] Available from: https://variohm.com/

[17] Available from: https://blog.endag.com/types_of_non-contact_vibration-sensors

[18] Available from: https://www.analog.com/media/en/training seminar/design-handbooks/Basic-linear-design/chapter3.pdf

[19] Introduction to temperature sensors: Thermistors, Thermocouples, RTDs, and Thermometers ICs September 16, 2022 by Nick Davis

[20] Thermostat: Definition, Functions, Components, Diagram, Working – Studentlesson

[21] Temperature Controllers - WIKA

[22] Available from: https://www.sensorland.com/howpage08

[23] Electronics Tutorial Home/Input/Output Devices/Temperature Sensors

[24] Courtesy of Danaher Industrial Controls Group - Process Automation, Measurement, & Sensing 1-800-884-4967

[25] Available from: https://www.instrucmentation.co.uk/non_contact-sensors_what_they_are_and_how_they_work/

[26] Comparing Contact and Non-Contact Temperature Sensors. Available from: azosensors.com

[27] How Sensors and Controllers Work Together | Gems Sensors

[28] A Comprehensive Guide to Non-Contact Sensors and Their Applications. Available from: mtiinstruments.com